The Living Oceans

The Living Oceans

by
Alec Laurie

Nature and Science Library
published for
Doubleday and Company, Inc.,
Garden City, New York

Editor   *Jonathan Elphick*

Designers   *Kevin Carver*
David Nash

Assistant   *Doug Sneddon*

Research   *Peggy Jones*
Naomi Narod
Judith Savage

Scientific Consultant   *D. J. Grove PhD.*

First published in the United States of America in 1973
by Doubleday and Company, Inc., Garden City, New York
in association with Aldus Books Limited, London

Library of Congress Catalog Card Number 70—178897
© Aldus Books Limited, London, 1972

Printed in Spain by Novograph, S.L., Madrid

CONTENTS

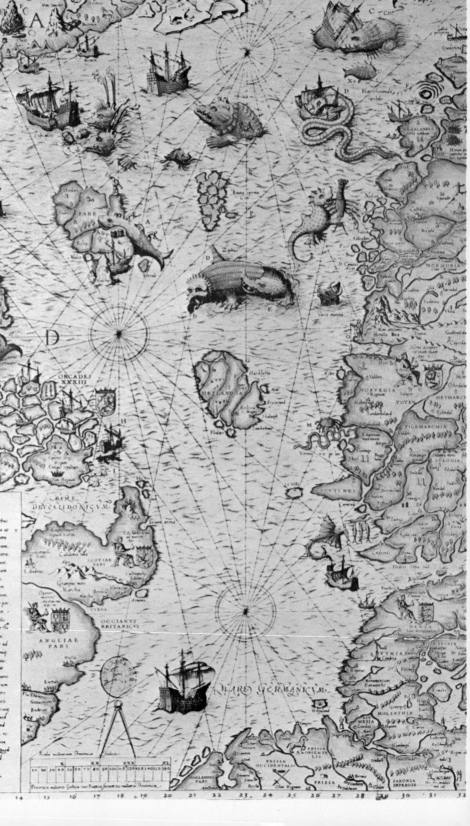

Introduction

Setting the Scene

When the astronaut Colonel Frank Borman toured Europe early in 1969, he said that, close up, the moon looked very much like certain parts of Texas—arid and monotonous. By contrast, the earth, as seen from *Apollo VIII* in moon orbit, was bright and colorful, and where the gaps in the cloud cover allowed a direct view of the earth's surface, the color was mainly blue. That the world's oceans are clearly seen at a distance of a quarter of a million miles emphasizes the fact that almost three quarters of the earth's surface is covered by water.

We can look at the *hydrosphere*—or watery part of the earth's surface—in several ways. In one view, it is no more than a thin film of salt water covering 71 percent of the planet's surface. In another, it is the main part of a gigantic desalinating (salt-removing) machine from which the sun's heat distills the fresh water that keeps all life going on the earth's landmasses. A third view likens the oceans to a vast highway, linking the great landmasses—and one that still provides the best medium in which to float and transport massive cargoes. Or one may look to the Bible and echo the Psalmist's idea of the sea as the vast habitat of countless creatures:

"The earth is full of thy riches. So is this great and wide sea, wherein are things creeping innumerable, both small and great beasts. There go the ships: there is that leviathan, whom thou hast made to play therein."

But the Psalmist had little idea, as he gazed out over the Mediterranean, of the enormous variety of life in the oceans. Since his day, ships have gone out all over the world, far beyond the Pillars of Hercules (now known as the Strait of Gibraltar), which for centuries was believed to be the end of the world. First explorers, then traders and pirates, and during the last hundred years scientists, have ventured everywhere that ships can go, from the equator to the poles. And at last we are beginning to know something about the oceans. Even so, we have thoroughly explored only about one 25-millionth of the total hydrosphere. We have done no more than skim the surface of the subject. But on the other hand we already have quite a useful collection of *samples* of what the sea contains, and we are gradually extending our knowledge of the oceans, from surface to sea floor.

A 16th-century Spanish chart of northern waters.
The detail here shows most of Norway, Iceland,
and the edge of Greenland. In these early maps
of the barely known oceans fact and fantasy—
icebergs and sea monsters—are shown together.

Simply to list all the plants and animals found in different regions of the sea would be dull. What is interesting is to take the oceans apart, and discover how differences in temperature, climate, and depth favor different kinds of plants and animals, how these live and breed, and how the total complex of life varies from place to place, and from one depth to another.

There is one very important difference between the pattern of life of the landmasses and that of the seas. On land all life is confined to a relatively thin layer, which is at most 200 feet thick. It is true that some birds and insects fly higher than this, and it is also true that some trees put down very deep-probing taproots in search of water. But in the main we can say that life on land begins about 3 feet below the soil's surface and ends at the highest treetops (or, where there are no trees, at a much lower level).

The pattern of life in the sea is quite different. There are shallow seas as well as deep oceans, but the *average* depth of salt water on this planet is 12,460 feet (2·4 miles), about the height of Mount Fujiyama in Japan and not so far off the height of Grand Teton Mountain in Wyoming, U.S.A. And marine life, in one form or another, exists at *all* depths.

So when we talk about the life of the sea, we shall be dealing with a living space about 160 times greater than the 200-foot thick habitat of the landmasses. Our geography will be three-dimensional, and we shall be examining the life of the sea not only in terms of its ocean-to-ocean distribution, but also in terms of depth, from the sunlit water of the surface to the cold darkness of the deepest ocean floor.

In the world today we know of approximately 1 million different *species* of animals. Broadly speaking, two animals are of the same species if they can interbreed and produce fertile offspring. For example, a poodle and a Great Dane belong to the same species because they can mate successfully and produce fertile young. The difference in their appearance is due to selective breeding by man. An alligator and a crocodile, on the other hand, belong to separate species. Although they look so alike, they do not interbreed successfully.

Of the earth's million species, the number in the sea is about 160,000, or approximately 16 percent. This may seem a rather small proportion, but remember that we are talking about the number of species, not about the number of individual animals. And when we compare the number of species in the sea with that on land, a surprising fact emerges: at least 75 percent of all land species are insects, whereas there is only a handful of species of marine insects. If we take away all the insects from the 1 million grand total, we are left with only 250,000 species on land and sea together— 160,000 in the sea, 90,000 on land. The result is that—other than insects— 64 percent of animal species live in the sea and only 36 percent on land. We can go a little further, and say that about 3,000 species live in the free water masses of the sea, and the remaining 157,000 on the seabed. Obviously we cannot attempt to deal with all of these. Instead we shall try to describe what lives where, and why, by dealing with some of the most important species.

Nautical Miles, Kilometers equivalents

Nautical Miles	Kilometers
1	1.9
2	3.7
3	5.6
4	7.4
5	9.3
6	11.1
7	13.0
8	14.8
9	16.7
10	18.5
20	37
30	56
40	74
50	93
60	111
70	130
80	148
90	167
100	185

Fahrenheit – Centigrade

Fahrenheit	Centigrade or Celsius
104	40.0
102	38.9
100	37.8
98	36.7
96	35.6
94	34.4
92	33.3
90	32.2
88	31.1
86	30.0
84	28.9
82	27.8
80	26.7
78	25.6
76	24.4
74	23.3
72	22.2
70	21.1
68	20.0
66	18.9
64	17.8
62	16.7
60	15.6
56	14.4
52	13.3
50	10.0
48	8.9
44	6.7
42	5.6
40	4.4
38	3.3
36	2.2
34	+1.1
32	0.0
30	−1.1
28	2.2
26	3.3
24	4.4
22	5.6
20	6.7
18	7.8
16	8.9
14	10.0
12	11.1
10	12.2
8	13.3
6	14.4
4	15.6
2	16.7
0	−17.8

Fathoms, Meters, Feet equivalents

Fathoms	Meters	Feet	Fathoms	Meters	Feet
1	1.8	6	500	914.4	3000
5	9.1	30	600	1097.3	3600
10	18.3	60	700	1280.2	4200
20	36.6	120	800	1463.0	4800
30	54.9	180	900	1645.9	5400
40	73.2	240	1000	1828.8	6000
50	91.4	300	1100	2011.7	6600
60	109.7	360	1200	2194.6	7200
70	128.0	420	1300	2377.4	7800
80	146.3	480	1400	2560.3	8400
90	164.6	540	1500	2743.2	9000
100	182.9	600	1600	2926.1	9600
110	201.2	660	1700	3109.0	10200
120	219.5	720	1800	3291.8	10800
140	256.0	840	1900	3474.7	11400
160	292.6	960	2000	3657.6	12000
180	329.2	1080	3000	5486.4	18000
250	457.2	1500	4000	7315.2	24000
300	548.6	1800	5000	9144.0	30000
400	731.5	2400	6000	10972.8	36000

Measurements taken of the oceans are often expressed differently from land measurements. These tables show conversions into land dimensions of nautical miles, used for distances, and fathoms, used for depths. Temperatures are also shown.

I

The Great and Wide Sea

What makes up the environment in which those "things creeping innumerable" live and move? In any natural community of living things it is the green plants that are the first or *primary producers* of energy. All life on land springs from the sun's energy. Green plants are able to use this energy, in the process of *photosynthesis*, to convert water and carbon dioxide into glucose. This is the primary source of food, which can be converted by living things into the energy needed for all life processes. A rabbit nibbling away at tender plant shoots is an example of a *primary consumer* or herbivore. Some of the energy within the plant food passes to the rabbit. The rabbit, in its turn, may fall prey to a hungry weasel. Animals, such as the weasel, that eat primary consumers, are called *secondary consumers* or carnivores. But the story of who-eats-who may not end here—a hawk may pounce on the weasel, and so the energy is carried one stage further.

There are also complications in that some of the energy is often channeled off by *parasites* (organisms that live on or in other animals or plants and feed on them). Finally, the body-wastes of living animals and the remains of dead animals and plants are broken down by bacteria and fungi in the soil into simple nutrients that can then be taken up by the plants again. These simple nutrients are thus endlessly re-cycled. (The energy from the sun is not re-cycled—it leaks away at every stage and finally escapes into outer space as heat. This is why a constant intake of solar energy is vital to life.) Biologists speak of such processes as we have described above as *food chains*. But nature is rarely as simple as we have suggested. The different food chains in a community are usually interwoven into intricate patterns that biologists call *food webs*. Nevertheless, the idea of the food chain is a convenient way of breaking down a fantastically complicated system into simpler parts that can be more easily understood.

It is a curious fact that the idea of a marine *food chain* that starts, and has to to start, with plant matter is a very recent one. The concept of plants as the basic material of life on land was commonly accepted thousands of years ago. Ever since man turned from hunting to becoming a herdsman, and then a farmer, it must have been obvious to him that, without vegetation as a

Across the immense and restless expanse of the Pacific Ocean these waves have traveled to curl over onto a California beach. The shining water, reflecting a setting sun, holds a whole world of vigorous life beneath its bright surface.

basic source of food energy, other forms of life could not exist. This notion is beautifully expressed in the saying "all flesh is grass." But the idea that a food chain at sea had to start with plants was overlooked until long after naturalists had begun to investigate the small organisms that drift with the currents and are collectively known as *plankton* (Greek for wanderer).

As far back as 1828 an English amateur naturalist, Dr. J. Vaughan Thompson, invented the first plankton-net—a silk bag that he towed through the water to filter out and catch small animals. But it was not until 1848 that a Danish botanist, Anders Ørsted, discovered that plankton consists not only of small animals but also of microscopic plants. He realized the significance of what he had found: that the first link in the chain of life is the same in the sea as it is on land—the green plant.

The contrast lies in the fact that plant life in the open sea takes an entirely different form, and consists of microscopic plant cells. The size of these microscopic floating plants determines the size of the herbivores that eat them. On land, herbivores range in size from beetles to elephants, but marine herbivores are mostly very small and are especially adapted to seeking out and devouring their microscopic food.

The most arid deserts and the frozen ice caps bear witness to the fact that there can be no life without water. All living matter consists largely of water, and indeed our own bodies comprise 60 percent water. Moreover, water is the "conveyor belt" by which foodstuffs pass in solution into living cells, and

Above: Scientists examining plankton specimens on board the Challenger *during her cruise around the world (1872–76). The specimens gathered and observations taken on that voyage have formed the basis for the modern study of the oceans.*

Right: A diagram showing the carbon dioxide cycle in the oceans. Carbon, one of the basic chemicals of life, is used in the sea and replenished by the carbon dioxide of the atmosphere, which is constantly passing into solution at the surface.

by which all waste matter is eliminated. Most important of all, water is also one of the two raw materials—the other is carbon dioxide—that plants use in the process of photosynthesis to produce the sugar glucose.

The water in seas and oceans provides an environment relatively free from the seasonal changes that, on land, temporarily halt the growth of plants. There are no summer droughts, and even in the depths of winter, life flourishes under the pack ice of the polar seas. This means that marine life suffers fewer slow-downs than land life, especially land life in the colder climates. It also means that the plant life of the sea does not have to build up a store of energy in the form of concentrated food with which to give the next generation a start in life. On land both annual and perennial plants prepare a seed, stocked with food, that is usually dispersed in the autumn and lies dormant through the winter, awaiting the warmth of spring before it can start an independent existence. This is not so in the sea. There plants grow vigorously and reproduce whenever the conditions are right. Low temperature is, in itself, no hinderance to the active growth of marine plants or of the animals that feed on them. Strange as it may seem, the most productive seas are the coldest, while tropical waters are relatively barren.

L and plants get the carbon dioxide they must have for photosynthesis from the atmosphere, which contains 0.03 percent by volume of this gas. This small concentration of the gas is enough for our present density of plant

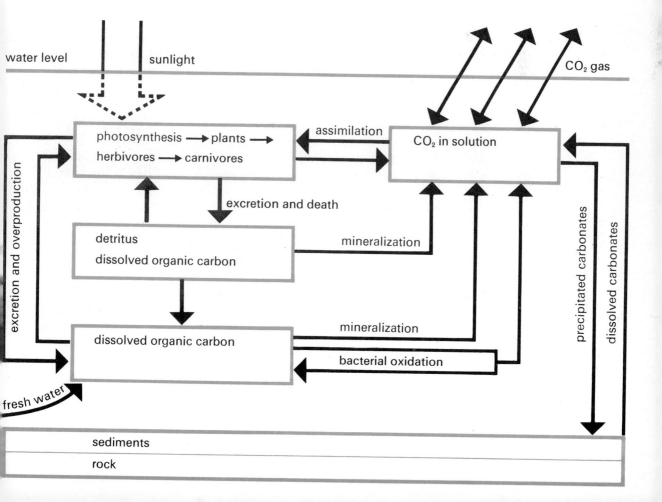

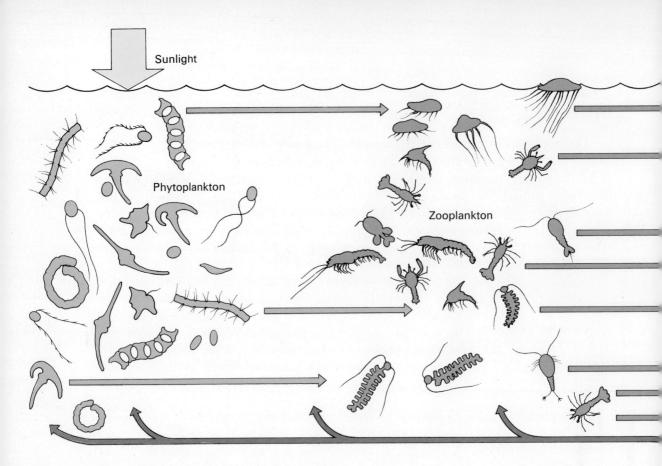

Sunlight

Phytoplankton

Zooplankton

growth. Nevertheless, plants would probably thrive better on more carbon dioxide. Nurserymen find that, by increasing the proportion of carbon dioxide in the greenhouse atmosphere, they can make tomatoes grow twice as fast as normally. This suggests that land plants growing in the open air can tolerate more carbon dioxide than they actually get. It is possible that, millions of years ago, during the Carboniferous period, when many millions of tons of carbon dioxide were absorbed from the air and ultimately converted into coal, there was much more of this gas in the atmosphere.

Clearly, healthy plant growth demands an ample supply of carbon dioxide. And in the waters of the sea there *is* an ample supply. We can best appreciate how abundant it is by comparing the amount of carbon dioxide that will dissolve in pure fresh water with the amount that is found in seawater. One liter of ice-cold distilled water, at normal atmospheric pressure, will dissolve less than 0.4 milliliters of carbon dioxide—that is, less than four ten-thousandths of a liter. One liter of seawater contains between about 90 and 150 times as much—the actual figures range from 34 to 56 milliliters.

The abundance of carbon dioxide in seawater is absolutely vital. Land plants rarely go short of carbon dioxide because, even on apparently windless days, slight changes in temperature and pressure make sure that there is at least some movement of the air around the parts of the leaf surface that absorb the gas. In contrast, the microscopic plants of the sea are free-floating. They move with the surface currents and at the same speed, so that,

Dominant oceanic marine food web, showing the pattern by which the energy of the sun is transferred by the food chain to the very bottom of the ocean. The phytoplankton, absorbing the sunlight in the surface layers of open water, form the basis of the chain. They are in turn fertilized by the typically coastal upwelling, shown by the long arrow at the left, which carries the nutrients produced by bacterial decomposition of organic materials that have drifted to the ocean floor.

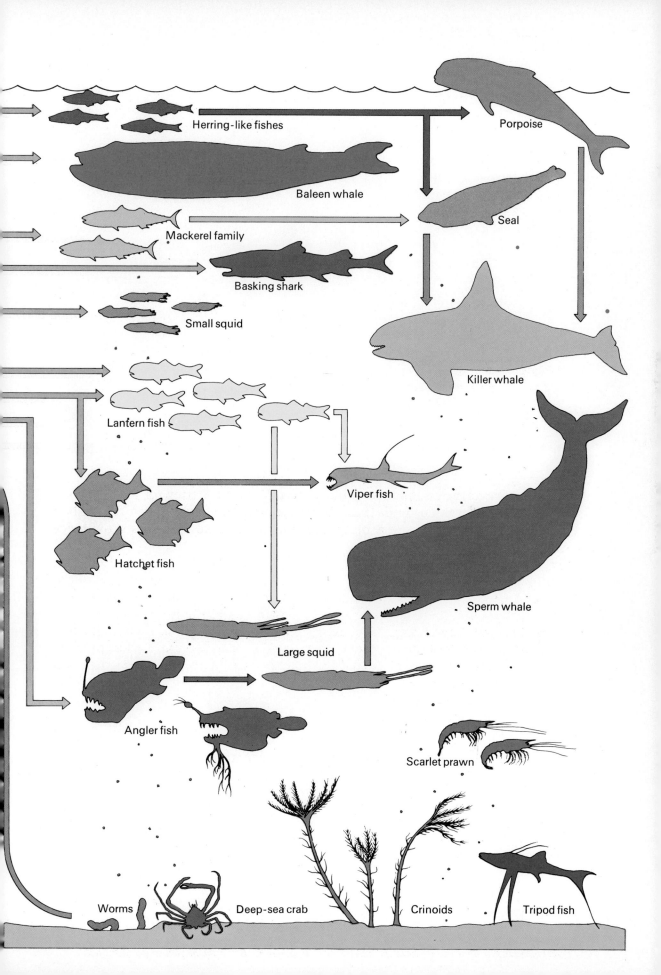

Herring-like fishes

Porpoise

Baleen whale

Seal

Mackerel family

Basking shark

Killer whale

Small squid

Lantern fish

Viper fish

Hatchet fish

Sperm whale

Large squid

Angler fish

Scarlet prawn

Worms

Deep-sea crab

Crinoids

Tripod fish

although they are mobile, they stay for a long while in the same place relative to the water surrounding them. So only the high concentration of carbon dioxide in seawater saves the plants from literally running out of gas. There may be, and often are, local shortages of other raw materials, which can prevent or slow down growth and reproduction of plants and animals— shortages of the kind that biologists call *limiting factors*. But lack of carbon dioxide is never a limiting factor at sea.

The earth's atmosphere contains far more oxygen than carbon dioxide— 20.99 percent by volume. But oxygen is far less soluble in water. So although there is so much of it pressing down on the earth's surface, a liter of ice-cold seawater will dissolve only 8 milliliters of oxygen. A liter of pure fresh water will actually dissolve more.

The sea, then, contains a strictly limited amount of oxygen that animals need for respiration. There is enough to provide the energy for the chemical changes that accompany growth, and enough for the movements that marine animals have to make in order to catch their food. But there is not enough oxygen available to enable a marine animal to indulge in prolonged bursts of speed, like a racehorse or an Olympic long-distance runner. Indeed, the only animals in the sea that are capable of high-speed exertion for more than a few minutes are the whales, seals, and other sea mammals, which are independent of dissolved oxygen because they come to the surface at intervals and breathe in the oxygen-rich air.

When we talk about the amount of oxygen available in a given quantity of water, we have to bear in mind the fact that living things cannot make use of all of it. Fish, for instance, extract oxygen by their gills, but the efficiency of their extraction apparatus is never 100 percent. The eel and the salmon can both extract up to 80 percent of the oxygen in a given volume of water, but the rate for most other fish is significantly less. Other marine animals have different methods of extracting oxygen, but whatever the method, the extraction process itself requires work, and therefore consumes oxygen. This applies equally to air-breathing land animals, of course, but the atmosphere is much richer in oxygen than seawater, so the work that land animals must do is less in proportion to the amount of oxygen extracted.

Whether an animal lives in the sea or on land, then, its capacity for sustained physical activity is limited by the amount of oxygen it can use. Because seawater contains a lower concentration of oxygen than does the atmosphere, the tempo of life in the oceans is slower than that on land. This does not mean that marine animals lack sufficient oxygen: it does mean that they have had to adapt their pace of living to their environment. Finally, oxygen is most plentiful in the surface layers of the sea, partly because these waters are in contact with the atmosphere, and partly because it is in this region that the process of photosynthesis by the microscopic plant plankton provides additional supplies of oxygen. As we go deeper, the oxygen is gradually used up by deep-sea animals, but even in the deepest known abyss (the 37,780 foot depth recorded in 1962 in the Mindanao Trench in the Pacific Ocean near the Philippines) there is still enough oxygen to support life, though at a very slow tempo. The reason why all the waters of the oceans contain oxygen is that there is a continual, though slow, circulation of water from the surface to the greatest depths. One of the few exceptions to this rule up to now occurs in the Black Sea, where the surface layer of fresh water is so much lighter than the salty water below that it floats on top without ever mixing.

The fact that there is less oxygen in water than in the air might suggest that marine creatures are at a disadvantage compared with animals on land. But life in the sea has its own advantages. In fact it is safe to say that, at any depth, a marine organism uses up less energy than a land animal in just staying alive and moving around. We can test this from our own experience. When we are lying flat on our backs with every muscle relaxed, we are not using up much energy, but when we are standing up, or even sitting, we are battling against the force of gravity. Many of our muscles are tensed all the time, otherwise we would fall down. We can never escape the force of gravity, although there are two ways in which we can appear to do so. These are by going into orbit around the Earth in a space capsule (a

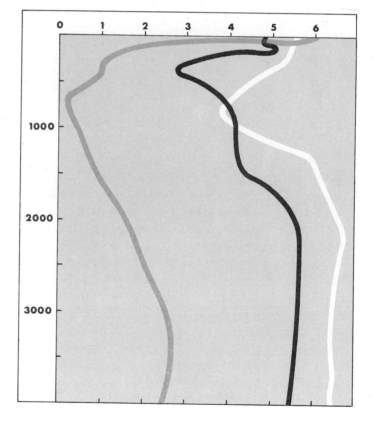

The vertical distribution of oxygen in different parts of the oceans. The figures along the top indicate the concentration of oxygen in milligrams per liter. Those on the vertical scale represent the depth of the water in meters. The blue line shows the distribution south of California, the black line in the eastern part of the south Atlantic, and the white line in the Gulf Stream. As the vertical diffusion of oxygen from the surface waters down to the depths is very slow, the particular pattern of the movement of the layers of water is most important in bringing oxygen to the deeper waters.

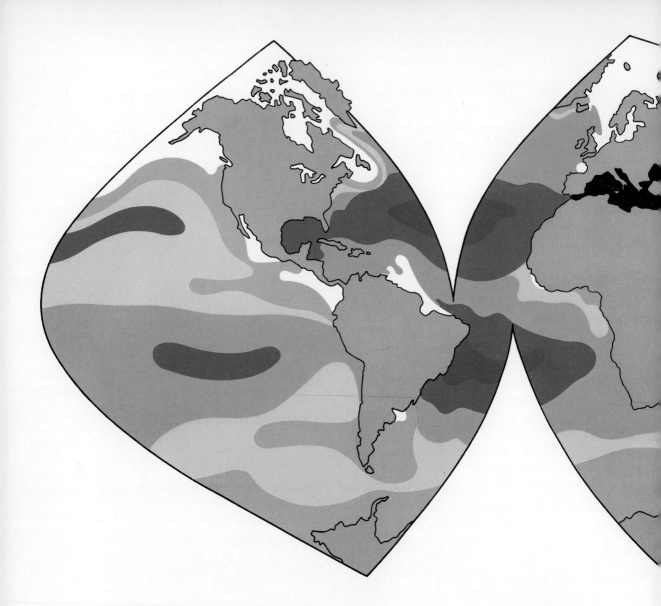

strictly minority pastime), and by swimming in the sea. In the latter case, we exchange the atmosphere, which gives our bodies very little support, for a medium that is over 800 times more dense—almost as dense, in fact, as our own bodies. Most of us, unless we are very fat, have some difficulty in floating in fresh water, but we can float easily in the sea. This is because the density of seawater is about 3 percent greater than that of fresh water at the same temperature. The densities of marine plants and animals are very nearly the same as that of the water in which they live. Few marine organisms habitually float on the surface, but there is every advantage in living in a medium that supports them so that they do not have to make a continual effort to prevent themselves from either sinking or rising.

Very few organisms are adjusted *precisely* to the density of seawater. There are, however, various ways in which they can alter or "trim" their buoyancy so as to remain effortlessly suspended in the water.

above 40

37-38

36-37

35-36

34-35

below 34

below 33

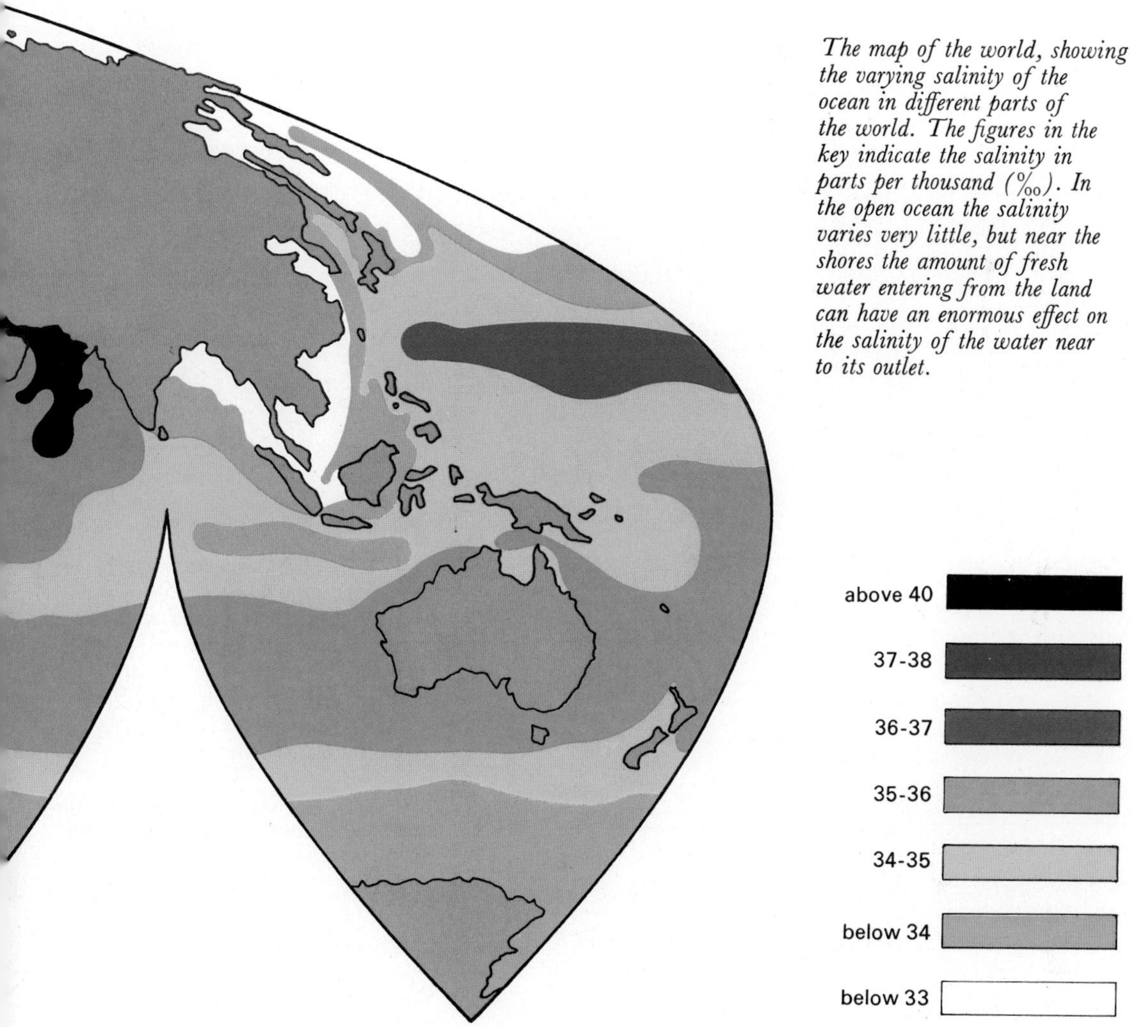

The concentration of salt in seawater is called its *salinity*, and this is always measured in parts per thousand by weight. A salinity of 30 parts per thousand by weight is written as 30‰, one of 40 parts per thousand by weight is written as 40‰, and so on.

A very rough figure for the average salinity of the open sea is 35‰, which is about five times the salinity of human blood. Salinity, however, varies from one part of the world to another. For instance, a nearly landlocked tropical sea with no rivers flowing into it, such as the Red Sea, may have a salinity of 40‰, because of the high rate of evaporation from its surface. At the other extreme is the Baltic Sea, with numerous rivers flowing into it and only a moderate evaporation rate. This results in a salinity of less than 29‰, and it can at times fall as low as 1‰ at the surface.

The Mediterranean is an interesting example of what can happen when salinities are drastically altered by local conditions. In summer the weather

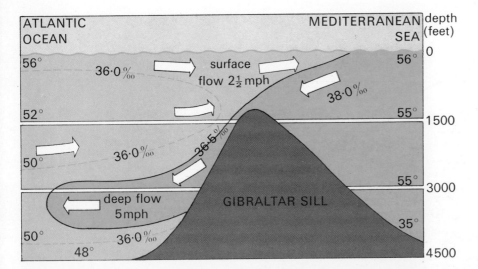

ATLANTIC OCEAN | MEDITERRANEAN SEA | depth (feet)

56°   36·0 ‰   surface flow 2½ mph   56°   0

38·0 ‰

52°   55°   1500

50°   36·0 ‰   36·5 ‰   55°

deep flow 5 mph   GIBRALTAR SILL   35°   3000

50°   36·0 ‰   4500

48°

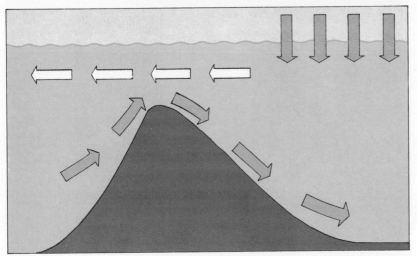

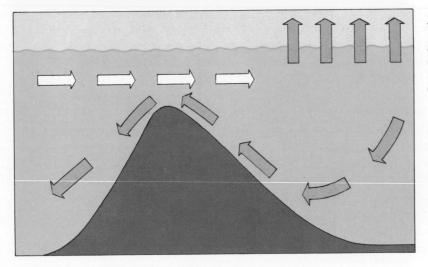

Top: The flow of water through the Strait of Gibraltar. The dense, highly saline water of the enclosed Mediterranean, shown here in dark blue, flows along the bottom of the strait. Above it, the waters of the Atlantic Ocean flow into the Mediterranean. The outflow of the dense salty water is more rapid, but lower in total volume than the inflow at the surface.

Middle: The pattern of circulation in adjacent seas when one sea has an excess inflow of fresh water (indicated by the gray arrows). This produces a flow of low salinity water at the surface (white arrows), while high salinity water moves in along the bottom (blue arrows). An example of such a sea is the Baltic.

Bottom: Circulation when one of two adjacent seas has a greater rate of evaporation (indicated by the gray arrows). This then creates denser high salinity water (blue arrows) that flows out under the inflow of lower salinity water (white arrows). The Mediterranean is an example.

in the Mediterranean basin is hot and dry, and this leads to heavy evaporation. Moreover, the rivers that flow into the Mediterranean do not bring in enough fresh water to compensate fully for this evaporation. If its outlet into the Atlantic through the Strait of Gibraltar were blocked, the level of the Mediterranean would therefore fall by perhaps as much as three feet per year. As it is, because of the high evaporation rate in the Mediterranean region, water there has a higher year-round salinity (36-38‰) than the water of the Atlantic (34-35‰), and is therefore heavier. Also, in the winter it cools and becomes heavier still. As a result of all this, Atlantic water enters the Mediterranean as a surface current through the Strait of Gibraltar, while the heavier Mediterranean water flows out into the Atlantic underneath. Because of the evaporation loss, the inflow is greater than the outflow.

Besides salt, seawater also contains a large number of other chemical compounds—often in very small quantities that have to be expressed in milligrams or fractions of milligrams per kilogram of water. Some of them are essential to life in the sea. Nitrates and phosphates, for instance, are just as vital to the growth of marine plants as they are to that of plants on land, while the elements calcium and silicon, found in other compounds, form parts of their skeletons. Phosphorus (in phosphates) is also an essential component of the skeletons, bodies, and cell fluids of marine animals. Iron, copper, and vanadium, found in yet other compounds dissolved in seawater, have their place in a variety of blood pigments that carry oxygen to the tissues (like hemoglobin in human blood).

Because plant life is the basis of all other life in the sea, just as on land, it is clear that a steady supply of nitrates, phosphates, calcium, and silicon is essential near the surface, where marine plants grow. It is also clear that, since plant life grows and is eaten, or dies and sinks to the bottom, there is a constant drain on these substances away from the surface, where they are most needed. If they are not replaced, plant production slows down and may even cease altogether. But the nutrients are replaced because there is a constant *vertical* circulation that brings up fresh supplies of essential elements from the lower depths. The word "vertical" is emphasized to distinguish this form of circulation from that of the surface currents, caused mainly by winds in the open seas and by tides in shallow waters.

We have seen how the difference in density between the water of the Mediterranean and that of the Atlantic results in an exchange of water through the Strait of Gibraltar. The same sort of movement takes place all over the world, and all the waters of the oceans are in constant slow circulation. Deep-water currents move only slowly, a typical speed of flow being between one-half and one mile per month.

The patterns of deep currents, and their effect on surface waters, are extremely complicated. One example of oceanic circulation will show the enormous scale on which the currents move. The diagram on pages 24-25

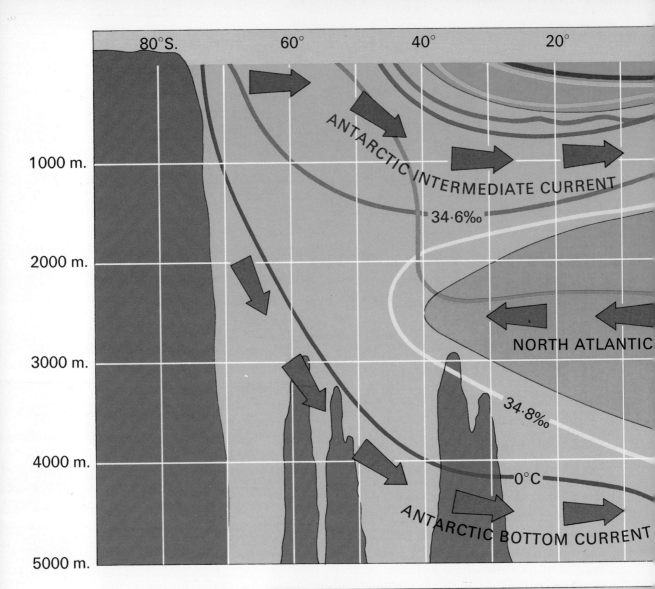

80°S.   60°   40°   20°

1000 m.

ANTARCTIC INTERMEDIATE CURRENT

34·6‰

2000 m.

3000 m.

NORTH ATLANTIC

34·8‰

4000 m.

0°C

ANTARCTIC BOTTOM CURRENT

5000 m.

Above: A diagram showing a simplified cross section of the Atlantic Ocean, with the ocean currents that move back and forth across the great distance. The colored lines join all points having the same salinity (‰) or temperature (°C).

Right: A diagram of the ocean layers, with the names marine biologists use to describe them.

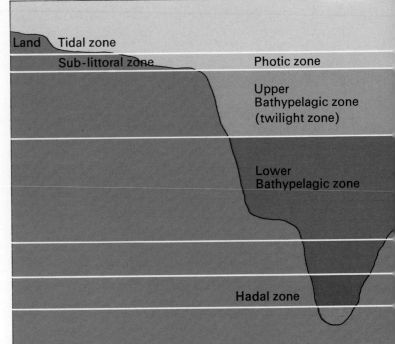

Land   Tidal zone

Sub-littoral zone   Photic zone

Upper
Bathypelagic zone
(twilight zone)

Lower
Bathypelagic zone

Hadal zone

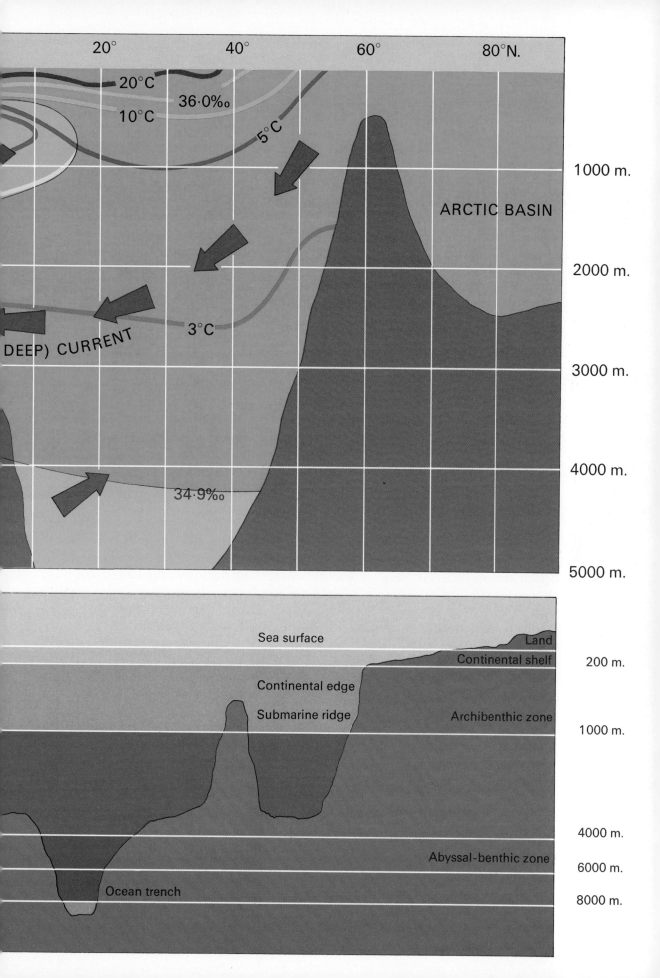

20°   40°   60°   80°N.

20°C
36·0‰
10°C
5°C

1000 m.

ARCTIC BASIN

2000 m.

3°C

DEEP) CURRENT

3000 m.

4000 m.

34·9‰

5000 m.

Sea surface   Land
                                    Continental shelf   200 m.

Continental edge
Submarine ridge   Archibenthic zone
                                                                      1000 m.

                                                                      4000 m.

                                    Abyssal-benthic zone
                                                                      6000 m.

Ocean trench   8000 m.

shows a vertical cross-section through the Atlantic, from the edge of the Antarctic continent to the North Pole (a distance of 11,520 miles). Between latitudes 25°s and about 44°N there is an *isohaline*—an imaginary line joining points having the same salinity. This particular isohaline is marked 36‰. This water has a high salinity, yet it is on top. The reason is that most of this water lies in the tropics. Its temperature is therefore high, and this high temperature reduces its density. (Its salinity is high because of the evaporation caused by a tropical climate.) High salinity increases the density, but not as much as high temperature reduces it. Therefore this layer floats on top of the north-flowing Antarctic Intermediate Current, which has a slightly lower salinity but a much lower temperature. It also floats on top of the south-flowing North Atlantic Deep Current, which also has a slightly lower salinity and an even lower temperature.

Yet another current, called the Antarctic Bottom Current, with a salinity of 36.7‰ and temperature of 32°F, flows northward and gradually mixes with the south-bound North Atlantic Deep Current. It may seem strange that the Antarctic Bottom Current, which comes from a cold region where there is little or no evaporation, should have the highest salinity of all. The reason is that when the sea freezes during the southern winter, the pack ice that forms on the surface consists of almost salt-free water, and the very cold water below the ice becomes loaded with extra salt, and therefore sinks.

We have talked about water and carbon dioxide, the raw materials of photosynthesis, and we can, for the moment, take for granted the presence of chlorophyll—which is essential for photosynthesis—in marine plants. This leaves us with the energy needed for the process—sunlight. There is a very considerable variation in the amount of light that penetrates the sea's surface, and there is also a big difference in the depth to which it penetrates in different regions. In the tropics, for instance, the sun at noon is almost vertical, so the amount of light reflected from the surface is very small. This means that in clear, tropical, ocean waters the light penetration is greatest. Yet, strange as it may seem, this is not where we find the greatest amount of marine plant life. The reason is that there is very little vertical circulation. Because the surface water is so warm, it is not very dense, and tends to *stay* at the surface. It does not mix to any great extent with the heavier, nutrient-bearing water below. The major circulations of deep waters are caused mainly by density differences, and because these differences are caused by a combination of temperature and salinity, this kind of circulation is known as *thermohaline*, (from the Greek *therme* =heat, *halos*=salt).

In high latitudes, and most of all in the Arctic and Antarctic oceans, the summer sun is low in the sky. Thus a great deal of light is reflected from the sea's surface, and very little penetrates. Nevertheless, it is in these regions that photosynthesis is greatest in the summer, partly because the days are longer, but mainly because these regions are rich in nutrients. So photosynthesis in the sea does not depend on bright sunlight.

In general, plant growth is restricted to the top 250 feet of water. In shallow coastal waters the depth limit is often considerably less than this because of the amount of sediment stirred up at the bottom by tidal streams and wave action, not to mention the mud that is brought down in river water. All these factors hinder light penetration. Even the plants themselves, multiplying rapidly in fertile waters, add to the general cloudiness, or *turbidity*, and reduce the depth of the zone in which photosynthesis takes place. Nevertheless, within this thin surface layer—a mere film compared with the vast depth of water beneath—grow all the plants on which every other form of marine life depends. This plant-producing surface layer is called the *euphotic* (well-lighted) zone, to distinguish it from the water immediately below, called the *dysphotic* (ill-lighted) zone, in which photosynthesis is meager or entirely absent. The two zones together extend to a depth of about 650 feet, which is generally the limit of effective penetration of light. This depth may fluctuate, however, depending on the clearness of the water and the height of the sun.

If we watch the setting sun we become aware of two things. One is that it steadily becomes less bright—we can gaze on it without strain. The other is the apparent change in color through deep yellow to red. By contrast, the sun at noon gives out what we call a "white" light, which consists of a fairly equal mixture of all the colors of the visible spectrum.

What is happening at sunset? The *quality* of sunlight that reaches the top of the atmosphere is no different from what it was at noon. But at sunset when the sun is low in the sky, the light has to take a much longer path through the atmosphere before it can reach us. The result is that those components of sunlight with the shortest wavelengths (violet, blue, and green) are stopped, while light of the longest wavelengths (yellow and red) get through to the earth's surface. So the setting sun looks to us like a fiery ball.

The sea, too, absorbs light of different colors, but its absorption rate is much greater, because water is a denser material than air. Moreover, seawater absorbs the different wavelengths of light in the opposite order: the red and yellow parts of the spectrum get absorbed in the top few feet, whereas the blue, green, and violet parts travel farthest before being reduced to a point where even a sensitive photoelectric detector can no longer register them. The depth of penetration of light of different colors varies according to the turbidity of the sea, as shown in the diagrams overleaf.

This *selective absorption* of light by seawater has a marked effect on the color of marine animals. On land we know of many animals that have *protective coloration*—that is to say, they are colored in such a way that they blend into the background, and hence avoid detection by their enemies. In the deep sea there is no detailed background, as there is in a jungle on land. There is only light from above and darkness in every other direction.

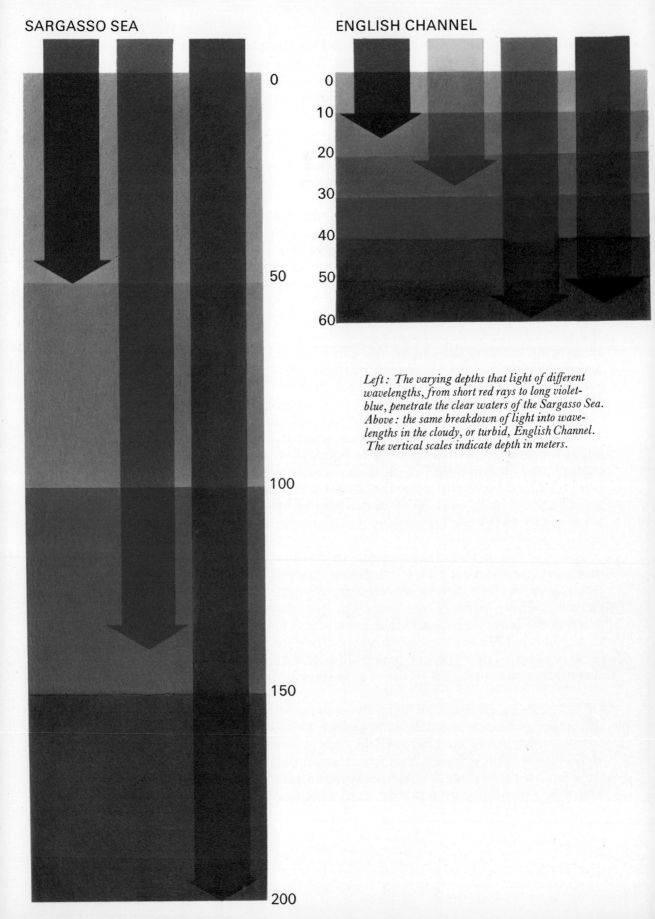

LIGHT PENETRATION

SARGASSO SEA

ENGLISH CHANNEL

0
10
20
30
40
50
60

0

50

100

150

200

Left : The varying depths that light of different wavelengths, from short red rays to long violet-blue, penetrate the clear waters of the Sargasso Sea. Above : the same breakdown of light into wavelengths in the cloudy, or turbid, English Channel. The vertical scales indicate depth in meters.

Accordingly, fish that live near the surface usually have dark backs, not easily seen from above, and lighter-colored bellies, which are hard to spot from below against the brightness above.

In deeper water, however, toward the limit of the photic zone, the light is narrowed down to blue, green, and violet. There the colors of animals tend to belong to the opposite end of the spectrum—yellow, orange, and red. To understand the advantage of this to the animal, we have to remember that what we call the color of an object consists of those parts of the spectrum that are not absorbed, but are reflected by its surface. Now, if an animal living near the bottom of the photic zone is colored red it will *appear* to be black against a dark background, because no red light has reached it, and therefore none can be reflected.

So in the sea, as on land, coloration often serves to make an animal inconspicuous, but the method by which this is brought about in the sea is very different from the method by which it is achieved on land. This is worth remembering when we look at pictures of gaily colored marine animals. Those that were photographed in their natural habitat near the surface of the sea are usually shown in their true colors. But the apparently gorgeous hues of many of the deep-sea animals shown in under-water photographs are misleading. Under natural conditions, the same creatures would appear either dark gray or black, and extremely difficult to spot in the gloom of their surroundings.

Above: This prawn's natural home is deep in the ocean, where only faint traces of violet-blue light penetrate. In this dark environment, the prawn's scarlet shell appears jet black, making it nearly invisible to any would-be predators.

2

The Restless Oceans

The movement of seawater has a profound effect on the distribution of marine life. On land, the barren regions and the fertile places remain fairly constant over periods of many years. But the areas of the sea where life is most densely concentrated are constantly shifting about. And their position in relation to land, and to the sea bottom, may change even from day to day.

Although there is much that oceanographers (scientists who study the oceans and ocean life) do not know about the circulations of the oceans, they do know that both the surface currents and those in the depths follow constant large-scale patterns. Surface currents have been studied over the years by mariners, who, by plotting their actual course against the course steered, have been able to calculate the influence of the currents. Today, the movements of deep waters, too, are being traced by oceanographers using a variety of scientific devices.

Several causes contribute to the major patterns of oceanic circulation. The most important are the effect of winds on the surface of the sea; the influence of the Earth's rotation; differences in the density of seawater caused by variations in temperature and salinity; and the shape of the landmasses and the seabed.

The relationship between the Earth's atmosphere and the oceans is a complex one. The Sun's rays penetrate the atmosphere and heat the sea, so that the sea acts as a vast storehouse of heat energy. The greatest amount of heat is stored in tropical waters where the heat from the almost vertical rays of the Sun is more intense than elsewhere. The air is warmed by the sea through contact at the surface, and is charged with water vapor as the seawater evaporates. This water vapor contains a large amount of potential energy in the form of *latent* (hidden) heat, which is released into the air as effective heat when the vapor condenses as rain, hail, or snow. Energy from the water vapor in the air generates winds, and some of these blow on the sea surface causing currents. At the same time they pick up more heat energy, making more winds, and so on.

The map on the next page shows the main winds that influence the course

A whirlpool. The ways that the restless oceans move vary from the broad sweeping surge of the major currents in mid-ocean to the flurries of eddies and whirlpools at the points where two great currents pass or collide with each other.

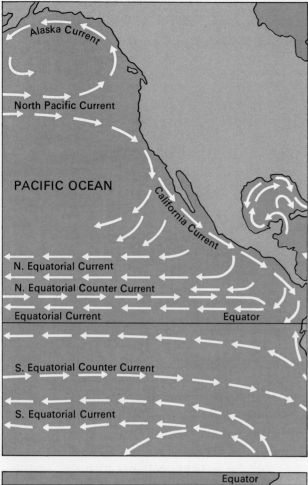

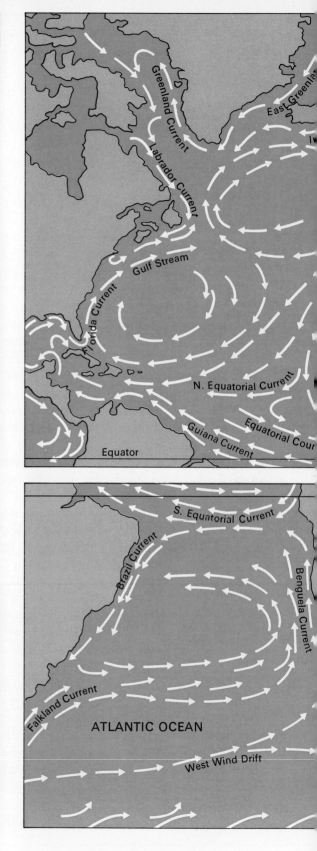

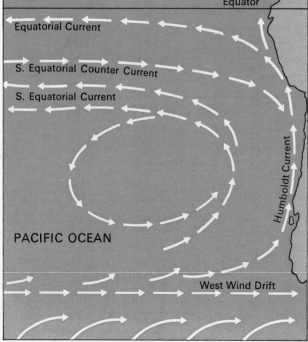

The major ocean currents of the world, showing the great gyres—the vast eddies that almost span the Atlantic and Pacific oceans both north and south of the equator. The gyres are created by the rotation of the earth, and kept turning by the wind systems. Several of the current patterns in the Atlantic and Pacific oceans are similar; for example, the warm Gulf Stream meets the cold Labrador Current, just as the warm Kuroshio meets the cold Oyashio, coming from the Bering Sea.

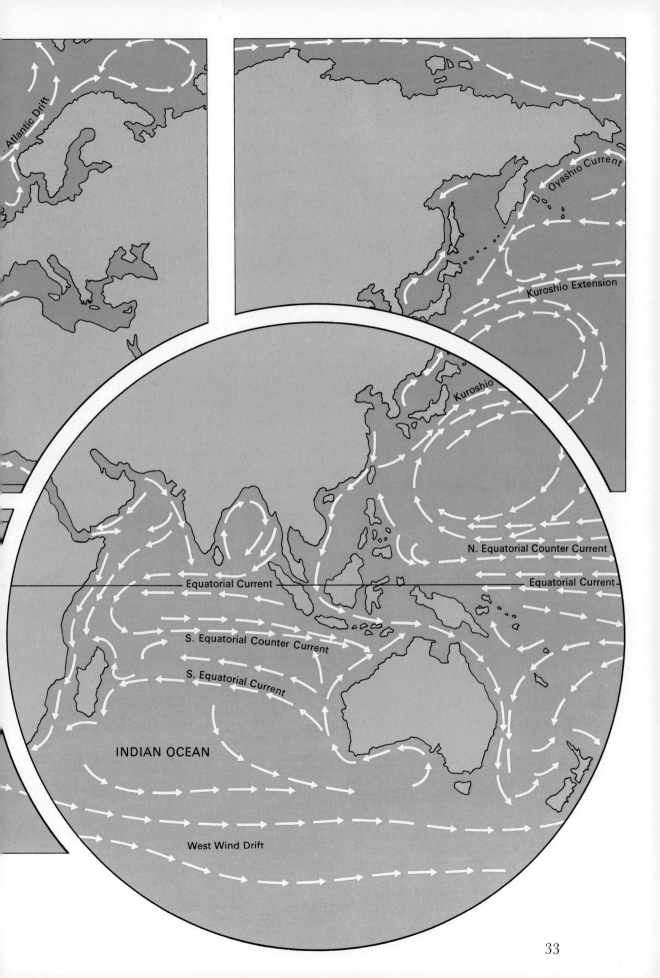

. Atlantic Drift

Oyashio Current

Kuroshio Extension

Kuroshio

N. Equatorial Counter Current

Equatorial Current

Equatorial Current

S. Equatorial Counter Current

S. Equatorial Current

INDIAN OCEAN

West Wind Drift

of surface currents. The most important are the Northeast and Southeast trade winds, which blow more or less constantly throughout the year, so that the Equatorial Currents in the Atlantic, Pacific, and Indian oceans move continually from east to west. To compensate for this one-way movement, there is a reverse flow of surface water in the Pacific and Indian oceans, forming the Equatorial Counter Current. A backflow such as this is hardly noticeable in the surface waters of the Atlantic. Between the Northeast and the Southeast trade winds is a calm area called the *doldrums,* where the winds are very weak. This area was a great hazard to seamen in the days of sail, and ships were often becalmed for weeks.

The map on the previous pages shows how the major currents flow in huge circular movements, called *gyres.* These gyres are formed partly because the great landmasses of South America, Africa, and Southeast Asia obstruct the flow of the Equatorial Currents, turning their course northward or southward. But the gyres continue, even where there is no obstructing landmass—clockwise in the Northern Hemisphere and counterclockwise in the Southern Hemisphere. The gyres are due to the Earth's rotation and are caused by what is known as the *Coriolis effect* (named after Gaspard Coriolis, a French mathematician who lived from 1792-1843 and who first fully described it).

The Coriolis effect is so important to an understanding of winds and currents that it is worth describing in some detail. Imagine that you are hovering over the Equator and facing north. You are in fact moving from west to east at the same speed as the Earth, as it spins on its axis. In other

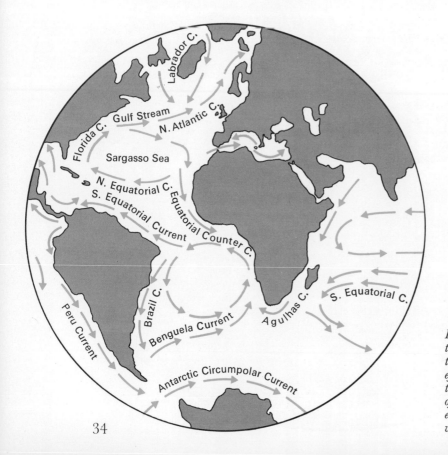

Left: The major currents of the Atlantic Ocean, showing the influence of the Coriolis effect, which can be seen in the basically clockwise turning of currents to the north of the equator, and the counterclockwise drift of currents to the south.

words you are hurtling to the right at something like 1,000 miles an hour.

If some force now sets you in motion so that you start to move northward, you will continue moving east at 1,000 miles an hour. But as you move northward, the ground or sea beneath you is not moving eastward as fast as it was at the Equator. The speed of rotation at the latitude of New York or Madrid (at about latitude 40°N) is about 780 miles per hour—220 mph less than your speed from west to east at the Equator. So by the time you reach latitude 40°N you will have traveled some distance eastward in relation to the Earth's surface—and an eastward turn as you move northward means a turn to the right. Strangely enough, if you were now set moving southward again, back toward the Equator, you would also end up to the right of your apparent destination—but now to the *west* of true south. This is because you would start off with, and *keep*, a west-to-east velocity of only 780 miles per hour, while the Earth's west-to-east speed would gradually increase up to 1000 mph by the time you reached the Equator.

It is therefore possible to state a general rule that bodies in free motion in the Northern Hemisphere are always diverted to the right. If we apply exactly the same reasoning as above to bodies in free motion in the Southern Hemisphere, we shall find that they are always diverted to the left. Bodies moving either east or west close to the Equator are not diverted. This can be seen from the diagrams below. If you look at the world currents map, you will see that the Coriolis effect does in fact work in this way on the winds.

Once the curving winds have set the surface water in motion, the currents also curve naturally. With this in mind we can examine in detail some of the world's major currents, looking first at the Northern Hemisphere.

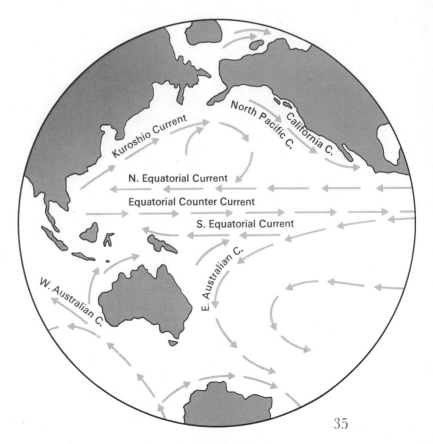

Right: The Coriolis effect on the currents of the Pacific Ocean. Here, where the equator crosses mostly open ocean, it is easiest to see how the Coriolis effect dwindles near the earth's midline, where currents move in a straight line.

35

In the Atlantic we can trace the course of one of the most powerful surface currents in the world—the Gulf Stream. Just south of the Equator the South Equatorial Current, driven by the Southeast Trade Wind, splits against the jutting landmass of Brazil. Part of it is deflected in a counter-clockwise direction by the Coriolis effect, becoming the Brazil Current, and part is turned northward and clockwise to cross the equator and join the North Equatorial Current.

Some of this northward flow enters the Gulf of Mexico as the Caribbean Current, and curves around until it finds an exit through the Straits of Florida between the tip of Florida and Cuba.

Most of this great stream of water then rejoins the Equatorial Current, and the combined flow—the Gulf Stream—travels up the Atlantic coast of North America. In the region of Newfoundland, part of the Gulf Stream carries on traveling in a northeasterly direction as the North Atlantic Current. This current conveys an enormous amount of heat to the north-eastern Atlantic. The rest of the Gulf Stream is turned east and south off Spain and down the coast of North Africa. There it rejoins the North Equatorial Current, so that a great clockwise gyre is formed around the Sargasso Sea, where there is hardly any noticeable surface movement.

The main body of the North Atlantic Deep Current, under the influence of the Westerlies (from the southwest), washes the coasts of western Europe from the Bay of Biscay right up to the North Cape of Norway and beyond. As a result, the Atlantic coastal waters of Europe are kept much warmer throughout the winter than those at similar latitudes on the other side of the Atlantic. For instance, the Russian port of Murmansk is kept ice-free all winter despite the fact that it is 124 miles north of the Arctic Circle, whereas the harbors of Labrador, at approximately the same latitude as the British Isles, are icebound from December to April. The economic importance of this to the countries of Western Europe is enormous: fishing and international sea trade are possible throughout the year.

The waters of the North Atlantic Current that flow around the British Isles and through the North Sea are known as the North Atlantic Drift, and take three main courses, as can be seen from the map. The massive inflow of water into the Arctic regions has to be balanced by an equal outflow. What happens is that the surface water south of Greenland is cooled and sinks, forming the Arctic Bottom Current. There is also an outflow of cooled surface water—the East Greenland and Labrador currents.

Off Newfoundland the cold Labrador Current meets with the westerly part of the Gulf Stream that travels up the east coast of North America. The mixing of the cold current and the warmer one provides ideal conditions for the spawning of cod. Indeed, the Grand Banks of Newfoundland have been fished continuously for cod (and also for its relative the haddock) since the 16th century. The cod is a fish that must have water at a temperature ranging from 39°F to 43°F for spawning. This condition is met in the Grand Banks area, at a latitude of 40-45°N, and cod thrive there. But on the eastern

Right: a flow map, showing the various streams, or filaments, of the Gulf Stream, each having a distinct temperature by which it can be tracked. The path of these small flows making up the large current varies weekly and even daily.

side of the Atlantic, the areas of commercial cod-fishing are much farther north—off the south coast of Iceland, at a latitude of about 63°N. Here we have a clear example of the far-reaching warming influence of the Gulf Stream and its importance to the distribution of marine life.

The other major ocean basin that reaches from the Equator to the Arctic is the North Pacific. There, too, the main direction of circulation is clockwise because of the Coriolis effect.

Right: A diagrammatic cross section of the Gulf Stream. North of the equator, the eastern water of the Sargasso Sea is warm and tends to expand and rise above the cooler, heavier water coming down the East Coast of North America. The eastern water flows toward the west (red arrows). But the Coriolis force diverts this current first toward the north, i.e. into the page (indicated by the red dots), and then toward the east. The dotted lines join points of equal water density.

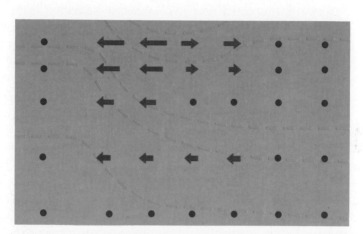

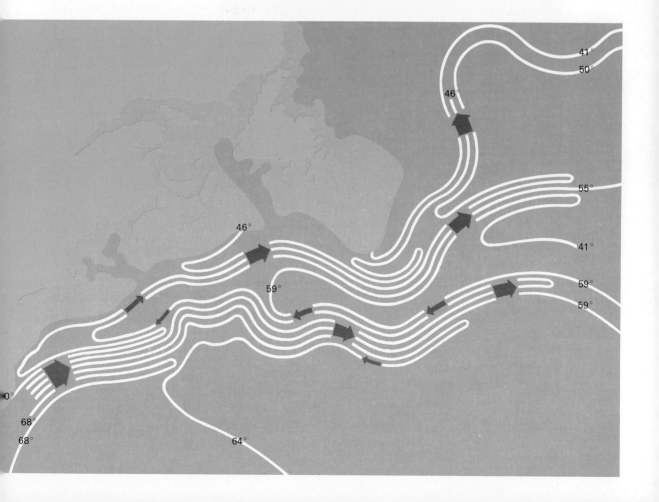

The North Equatorial Current turns northward from the Philippines and travels along the south coast of Japan, where it is called the Kuroshio Current. This current then turns east and flows across the North Pacific (as the North Pacific Current), gradually cooling until it reaches the west coast of North America. There part of it turns northward up the coast of Alaska to form the Alaska Current. The remainder flows south down the North American coast, where it is called the California Current.

The surface waters off the west coast of North America are much colder than those off the coasts of Europe at the same latitude. For instance, the waters off the south coast of Alaska have an annual temperature range of between 37° and 54°F—the same as the Norwegian Lofoten Islands, which are 558 miles nearer the North Pole. Farther south, the summer temperature outside San Francisco Bay (latitude 37°N) is almost the same as that at the mouth of the English Channel (latitude 50°N).

There is no consistent pattern of strong westerly winds in the North Pacific, as there is in the Atlantic, and the waters of the North Pacific Current are not driven into the Arctic Ocean. Water does, however, flow *out* of the Arctic Ocean, and is evident as a surface current moving down the west side of the Pacific toward North Japan, where it is called the Oyashio Current.

This, very briefly, is the major pattern of surface currents in the Northern Hemisphere. All the surface currents we have been discussing, with the exception of the East Greenland, Labrador and Oyashio currents, have their origins in the tropics. Their effect is mainly that of raising sea and air temperatures and of pushing the ice-free limit of northern water well into the Arctic regions. The currents themselves contribute little to the productivity of these areas because, coming from the tropics, they are poor in nutrients. It is only when these warm currents mix with colder and more fertile water— for instance, off the western coasts of Europe—that the coastal waters become really productive.

In the South Pacific there is a counterclockwise circulation of currents into and out of the Antarctic Ocean, similar to that in the South Atlantic. On the eastern side of the Pacific there is an area of great upwelling of fertile bottom water in the Peru Current. (This flow is sometimes also called the Humboldt Current, after Alexander von Humboldt, 1769-1859, the famous German explorer of South America). It used to be thought that the Peru Current was simply a section diverted from the world's greatest mass of moving water, the West Wind Drift, but oceanographers now know that the pattern is more complicated.

The vast quantity of water in the West Wind Drift is circulated by almost continuous winds from west to east around the continent of Antarctica. This water has an unobstructed flow south of Africa, Australia, and New Zealand, but the tip of South America sticks out farther south than the other land masses, so there the water meets an obstruction. When it comes up against this obstacle, part of it is able to pass through the Drake Passage between Cape Horn and the South Shetland Islands. The 370-mile-wide passage is too narrow for the amount of water building up to the west, and this bottle-neck is partly the cause of the rough and dangerous seas in the area. Some of the water that cannot get through the Drake Passage comes around the southern tip of South America and some is diverted as a spur of cold water up the west coast of Chile and Peru as far as the Equator, where it merges with the South Equatorial Current. This cold current does not follow the coast closely, but flows about 30 miles out to sea, and is therefore known as the Peru *Oceanic* Current. It does not arise from the highly fertile water of the Antarctic, but from that part of the West Wind Drift to the north, which has a relatively low fertility. So the waters of the Peru Oceanic Current do not support a heavy crop of marine life. Yet if we move toward the coast, we find the waters there teeming with sea creatures. Why is this so?

The western coastal waters of South America are very deep. There is little or no continental shelf, and soundings of 3,300 feet within 3 miles of the shore are quite common. The winds in this area are variable, but their general direction follows the coastline from south to north in Chile and from southeast to northwest in Peru. These winds, which blow parallel to the coast, move the surface water in an irregular way, and this water is replaced by the upwelling of deep water, which is colder and also remarkably rich in phosphates. The zones of upwelling can be easily spotted by the opaque green color of the water where there is a particularly heavy growth of *phytoplankton* (plant plankton). These green patches contrast strongly with the clear water of the Peru Oceanic Current.

Probably nowhere in the world, apart from the Antarctic Ocean, is there such a wealth of life as in this narrow coastal strip of water. From the beginning of the Peru Current just north of the Island of Chiloe (Chile) at latitude 42°s to the point where it leaves the coast of Cabo Blanco (Peru) at latitude 4°s, there is a marked difference in the types of animals found in the coastal waters. This is a result of the gradual change in the temperature of the surface waters from an average of 56°F in the south to 63°F at the extreme north. In the south the vast quantities of phytoplankton provide food for a heavy crop of crustaceans, fish, squids, and whalebone and sperm whales (the latter being squid eaters). In the north, off Peru, where the fertile zone reaches its greatest width (about 93 miles), the temperature is high enough for the small fish called anchovies to flourish.

Most small fish eat still smaller fish that in turn prey on zooplankton (tiny plankton animals), which eat the phytoplankton. But this is not the case with anchovies. Anchovies feed directly on phytoplankton, and so shorten the food chain. This is important, because each stage of the food chain from plant to small animal to larger animal is only about 10 percent effective. For example, in two links of a food chain it takes about 10 pounds of tiny shrimp-like copepods to produce 1 pound of fish. If, as with anchovies, the inter-mediate stage of small animals is by-passed, a given amount of phytoplankton can increase its yield of fish a hundredfold. The result is that off the coast of

Top: Anchovies in the hold of a fishing vessel off the Peru Coast. The nutrient-rich upwelling of the cold deep waters supports vast numbers of phytoplankton, on which enormous shoals of anchovies can feed directly.

Above: A thriving industry has developed in Peru, based almost entirely on anchovies. The catch is processed into fish meal for animal feed, shown here in sacks at a factory near Pisco, ready for export.

killed that at times the army has been called out to bury the piles of them that are washed up on the beaches. And with the disappearance of the fish, the seabirds, too, starve to death in their millions.

The Peru Current ends off Cabo Blanco, where its waters merge with the eastern edge of the Equatorial Counter Current. Together, these waters form the northern limit of the South Equatorial Current. There is an abrupt change of surface temperature where the two currents meet, and the colder water (about 63°F) dives beneath the tropical water (which has a temperature of between 72°F and 75°F). The meeting point is not sharply defined, but consists of a series of eddies and whirlpools. In them, green and blue water can be distinguished, and the characteristic muttering sound of two bodies of water in collision, called a *tide rip,* can be heard. The colder water does not yet run right under the tropical water. The two mix, and the South Equatorial Current is relatively cold for the first few hundred miles. The Galàpagos Islands, situated almost on the Equator, benefit from this cool and still-fertile water, because the blend of two currents, one from the Central American region and the other from Chile and Peru, brings a greater supply of nutrients than is usual in equatorial regions. This blending also results in a mixture of tropical and temperate animals that are *eurythermal*—that is, they can live over a wide range of temperatures.

I n the South Atlantic, as in the South Indian Ocean and the South Pacific, the main gyrations of the surface currents flow in a counterclockwise

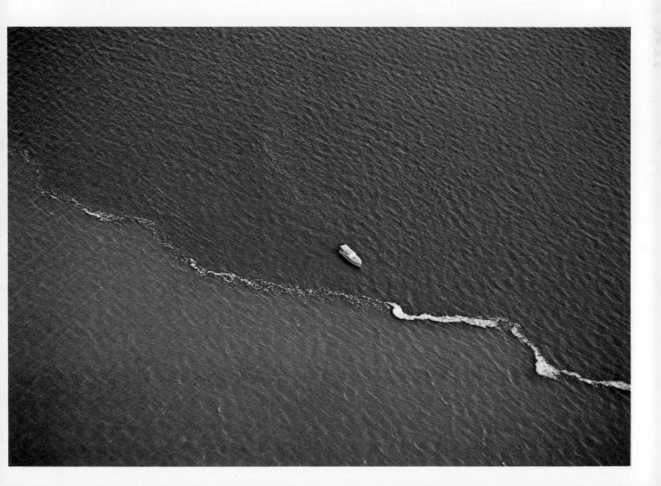

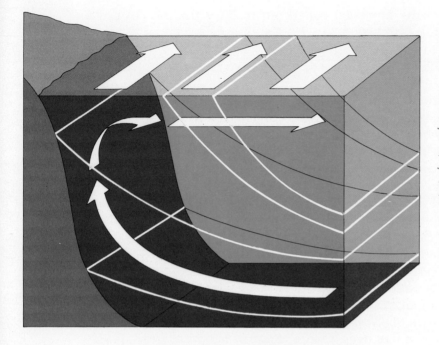

In coastal upwelling, as shown here, winds near a coast blow the surface waters along, and the deep cold water rises in replacement. This water from the depths brings with it a load of nutrients that have drifted down toward the ocean floor, and which then serve as rich fertilizer for the plant plankton. This means more food for the microscopic animal plankton feeding on the plant plankton, and therefore also for the larger animals that form the final stages of the food chain. Where this upwelling takes place, as in the Peru Current, ocean life is extraordinarily rich.

direction because of the action of the Coriolis force. Part of the body of water turned south at Brazil is further deflected to the east, and crosses the Atlantic between the latitudes of 30°s and 40°s until it meets the landmass of West Africa. There it turns north and travels up the coast as the Benguela Current, merging eventually with the Equatorial Current. The Benguela Current (named after the town of that name in Portuguese Angola) is also partly formed by the West Wind Drift. The northern part of the West Wind Drift is diverted toward the north by the Coriolis effect and cut off by the southern tip of Africa. Along the coast, north of Cape Town for a distance of about 745 miles to latitude 19°s, there is a marked upwelling of water, caused mainly by the Southeast Trade Wind blowing offshore.

The surface waters of this current are comparatively cold, with a mean temperature of 50°F—about 18°F less than at Rio de Janeiro, on the other side of the Atlantic. In the cool surface waters of the Benguela Current there is intense plankton activity, largely because of the constant renewal of nutrients by the upwelling of deeper water. Here we find conditions very similar to those in the fertile Peru Current. Occasionally, in the region of the Benguela Current, there is an outbreak of poisonous dinoflagellates, which results in the death of millions of fish. There are also heavy fish losses caused by lack of oxygen.

At Walvis Bay in South West Africa, for example, shoals of dead pilchards are frequently washed up on the beach. On investigation, these fish appear to die from lack of oxygen rather than from poisoning. It seems that the daytime oxygen production of the phytoplankton is at times not enough to provide for its own night-time respiration as well as the oxygen demands of the massive zooplankton and fish populations.

The shoreline of this area is densely populated with worms and mollusks. But offshore for about 17 miles out—beyond the range of the wave action that keeps the sandy bottom stirred up—the seabed is covered with a soft, dark green mud that gives off hydrogen sulfide gas and contains skeletons and corpses of animals. No life can exist there except anaerobic bacteria.

It seems that we can draw but one conclusion. If life in the upper waters is extraordinarily plentiful, the fall-out of dead matter is too great to be broken down, and too great for much of the nutrient to be re-cycled. This harms or kills most creatures that depend on oxygen.

L et us now take a look at the great expanse of ocean that encircles the frozen south polar continent. This ocean is called the Antarctic, or Southern Ocean. It is unique in that it completely surrounds a continent, without being broken by land.

The Antarctic surface waters are among the most productive in the world: they have an exceptionally high content of inorganic nutrients (nitrates and phosphates), and can therefore support huge quantities of phyto- and zoo-plankton. To a great extent it is the currents that are responsible for this abundance of life.

During the spring and summer the surface layer of the Antarctic waters is formed at the ice-edge from a mixture of the North Atlantic Deep Current and melting ice. (This melting ice is partly winter pack ice and partly ice from the Antarctic continent, and from icebergs that have broken off from it). The water is, of course, very cold, but it does not sink. This is because it is diluted by the fresh water from the ice, which reduces its salinity to less than 34‰ and renders it less dense than the more saline water beneath. Because of the great quantity of phytoplankton that is continually giving out oxygen during photosynthesis, this surface water contains a relatively high concentration of oxygen, particularly in summer.

During the winter, the surface water is still flowing northward, but it is diluted less by fresh water because there is much less ice melting. There is also less oxygen, because the phytoplankton has far less light for photosynthesis during the short days of the polar winter, and therefore gives off far less oxygen than during the summer. A line can be drawn on a map of the Antarctic Ocean, completely encircling the Antarctic continent, between 50°s and 60°s latitude. This line marks the region where surface waters flowing northward meet southward-flowing water, and it is called the Antarctic Convergence.

A t the Antarctic Convergence, the surface waters sink to form the Antarctic Intermediate Current, and continue to drift northward. During this drift the salinity of the water increases gradually to 34.9‰ as the current mixes with the water above and below. The amount of oxygen in

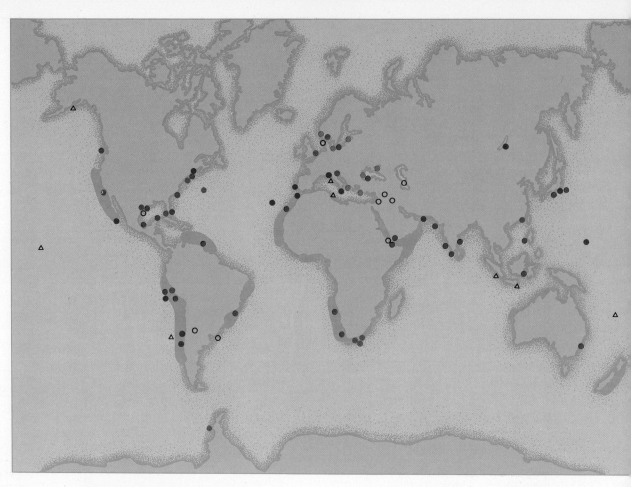

Reasons for death:

△   volcanic disturbances   ●   seaquakes

○   changes in salinity   ●   red tide

●   changes in temperature   ●   hydrogen sulfide

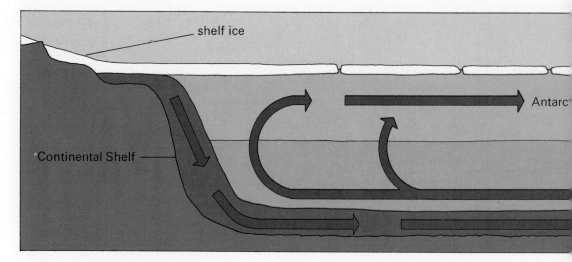

shelf ice

Antarc°

Continental Shelf

Top : A worldwide picture of the places where marine life has been killed by natural disasters. The toxic plankton organisms that cause red tides flourish where there are unusual increases in nutrients, coupled with calm, warm water.

Above : A profile of the Antarctic Ocean, showing upwelling of the warm deep nutrient-rich current from the north. It replaces the cold salty water, which slides to the bottom, and the surface water, diluted by fresh water from melting ice.

the water gradually diminishes as it is used by zooplankton and fish.

From measurements of its salinity and oxygen content, oceanographers have estimated that the Antarctic Intermediate Current takes at least seven years to travel the 4,700 miles from the Antarctic Convergence (at latitude 50°s) to about latitude 20°N. This means that the current travels at a speed of almost two miles per day.

There is another, more saline, current that originates in the Antarctic Ocean. This is the Antarctic Bottom Current. Oceanographers have mapped its course from the Antarctic ice-edge to latitude 40°N. This current starts beneath the pack ice. When ice is formed in the winter, salt is released from the freezing water, and the water below the ice therefore becomes more salty and denser. This highly saline water, cooled by the pack ice above, becomes even more dense and sinks. Between them, the Antarctic Intermediate and Bottom currents dominate the current patterns of the Antarctic Ocean.

The tremendous outflowing of water from the polar regions demands an inflow to make up for the loss. This is provided by the south-going current that flows between the Intermediate and Bottom waters. In southern waters it is often referred to as the Warm Deep Current (though the term "warm" is only relative). It is also called the North Atlantic Deep Current, because it can be traced throughout the entire Atlantic Ocean, from the cooled surface water that sinks off Greenland (where it is known as the Arctic Bottom Current) right down through the North and South Atlantic, to the point where it enters the Antarctic Ocean at an estimated rate of 918 million cubic feet of water per second.

Clearly, then, the patterns of circulation of deep waters throughout the world are highly complex, and oceanographers do not yet fully understand them. In the following chapters, however, we shall see something of their great significance to marine life.

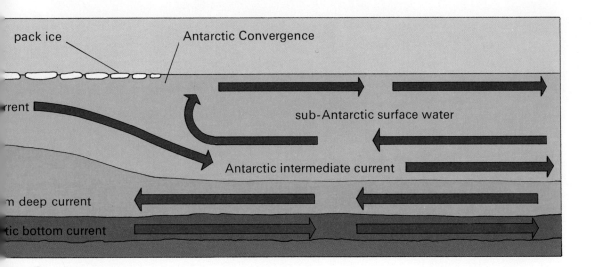

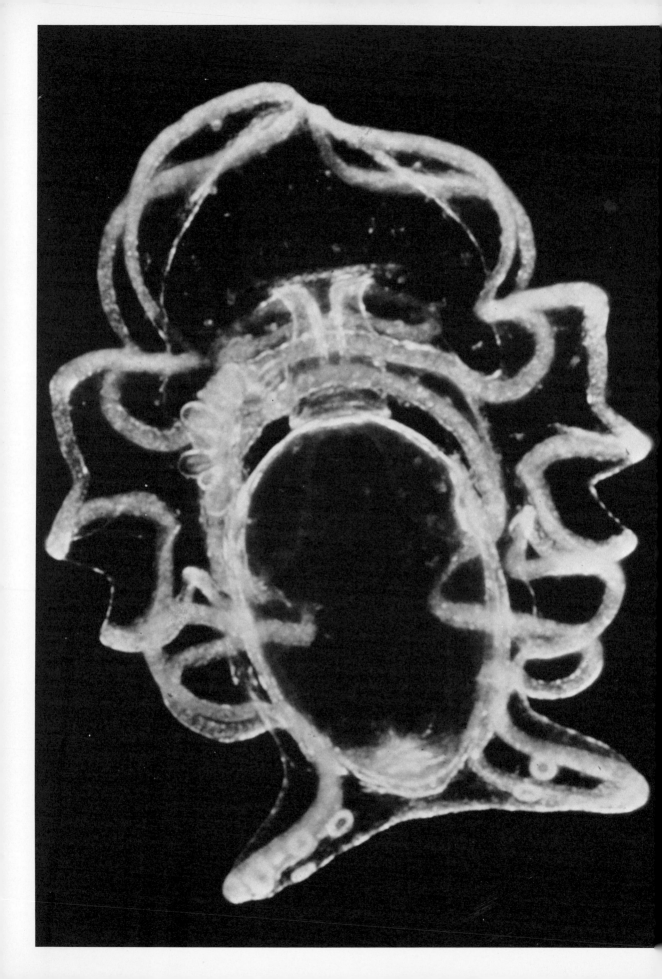

3

The Sea Drifters

The adjective we most frequently use to describe the sea is "blue." The word comes almost automatically to mind, particularly if we are thinking of the open oceans, or rather, *not* thinking of the shallow coastal waters. This is worth looking into, because the sea's color is a good indication of the amount of life in it. (We have to rule out coastal and shallow waters because in these areas the action of waves, tides, and rivers stirs up the bottom and makes the sea cloudy or *turbid*).

In the tropics, when the sun is shining, deep water is seen as dark blue with a touch of violet. When a cloud covers the sun, the color is very dark blue. Whether the sun is shining or not, however, blueness is the keynote, and with blueness there is transparency. If you stare down through the water for a few moments you get the feeling that you are at a great height, and if you happen not to have a good head for heights, the experience can be quite alarming. The writer well remembers going swimming 1,000 miles from land in the South Atlantic, where the depth of the ocean was 13,000 feet. The water was intensely blue and clear, and it was easy to feel panic at being so far out of one's depth. The only way to keep going and enjoy the swim was to hold on to the idea that one could just as easily drown in 13 feet of water as in 13,000 feet.

The blueness and clarity of deep water indicates a scarcity of life, because there is very little in the water to reflect light—except water itself. And since, as we have seen, it is the blue and violet rays that penetrate deepest, these colors are the ones that will in the end be reflected farthest by the water particles and return through the surface to give us the blue-violet "message." Just as one can tell at a glance by the yellowy-brown color of a landscape that it is a desert, so one can identify the deserts of the sea by their clear, deep blue coloration.

The color green is, of course, the indicator of vegetation, and this is just as true of the sea as it is of the land. So when we come across green water, we know that there is a great deal of plant life in the surface layers. Also, where there is a plentiful supply of vegetation there will also be a large population of animals, feeding on the plant matter.

A sea cucumber larva, one of the tiny animals that graze on the phytoplankton. It drifts as a member of the zooplankton only during the early stages of its life: it then becomes a slug-like creature crawling over shores and the seabed.

In examining the conditions in which marine organisms live the the word *plankton* has been used to describe those inhabitants of open seas that cannot swim against currents or tides. When looking at these tiny wanderers in more detail, we find that we can divide them into two main types: the first is the *phytoplankton* (Greek *phytos*—plant), consisting of drifting plants, and the second is the *zooplankton* (Greek *zoon*—animal), which consists of drifting animals. The phytoplankton will be considered first because plant matter is the basis of all marine life.

Most important and widespread among the phytoplankton are the *diatoms*. Ranging in size from 1/000 inch to 1/25 inch, these single-celled plants have an enormous variety of shape and structure, but they all possess one feature in common: their skeletons are made of *silica,* a sort of natural glass. The interior of the cell, with its chloroplast (containing the all-important chlorophyll, the pigment used in photosynthesis), is clearly visible under the microscope. The structure of diatoms may well seem very exotic, but in fact it is strictly practical. The elaborate skeleton is pierced with minute holes that provide contact between the cell and seawater. The array of small spikes causes a great deal of friction with the surrounding water, and helps prevent the diatom from sinking. This is important, for a diatom's very life depends on being able to stay in the euphotic zone. We even find that diatoms living in warm seas—where the viscosity (resistance to movement) of the water is much less than it is at low temperatures—have more elaborate skeletons than the cold-water species, which serve even more effectively to prevent sinking.

The question of "sink or swim" is vital. There are two things that help diatoms to stay near the surface. The first is that all of them have an extremely large surface area in proportion to their body weight. The second is that some, though not all, produce a small amount of fat, in the form of minute droplets of oil, inside the cell. Because oil is lighter than water, this provides buoyancy, and makes up for the high density of the diatom's glassy skeleton. It is probable that diatoms only just manage to stay near the surface long enough to reproduce by simple division—or to be eaten. What is certain is that when they die they lose their buoyancy and start to sink, so that there is steady rain of dead diatoms falling from the euphotic zone, like leaves during fall. Much of this fallout must supply food to animals that live in the deeper waters, but the skeletons of many of the larger diatoms reach the bottom. where they accumulate in such vast numbers that the mud of the ocean floor consists almost entirely of what is called *diatomaceous ooze.* This ooze has been accumulating for millions of years, and by taking bottom samples with a *corer* (a long hollow tube that can be driven into the seabed) it is possible to map the regions where there are most diatoms in the surface waters. It is no surprise to find that the whole of the Antarctic Continent is ringed with diatomaceous ooze, because the Antarctic waters are among the most fertile in the world. Also, it is interesting to see that the northern

Above right: A photomicrograph of delicate, pill-box-shaped diatoms. Single-celled plants, they are the most widespread of the phytoplankton. Right: All diatoms have skeletons of silica, but they exist in a great variety of intricate forms.

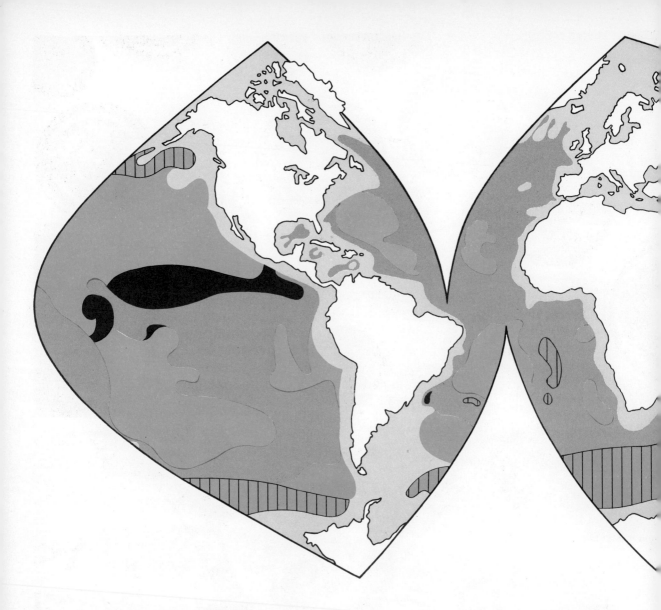

limit of ooze corresponds quite closely with the line of the Antarctic Convergence—the point at which the cold, fertile surface water sinks under the warmer and lighter sub-Antarctic water, which is much less fertile.

Skeletons of long-dead diatoms are not confined to the ocean bed. Deposits of *diatomaceous earth* are often found where a former seabed has been raised and has become dry land. Because the intricate skeletons are often perfectly preserved, many of them are recognizable as belonging to the same species as those that float in today's oceans. These fossil deposits, mostly dating from about 60 million years ago, are much in demand as a polishing agent, and as an ingredient of some toothpastes. And because of their very large surface area compared with their volume, they are particularly useful as an absorbent for nitroglycerin, a powerful, but highly sensitive explosive, which by itself is liable to explode spontaneously. The combination of diatomaceous earth and nitroglycerin is called *dynamite*, which is quite safe

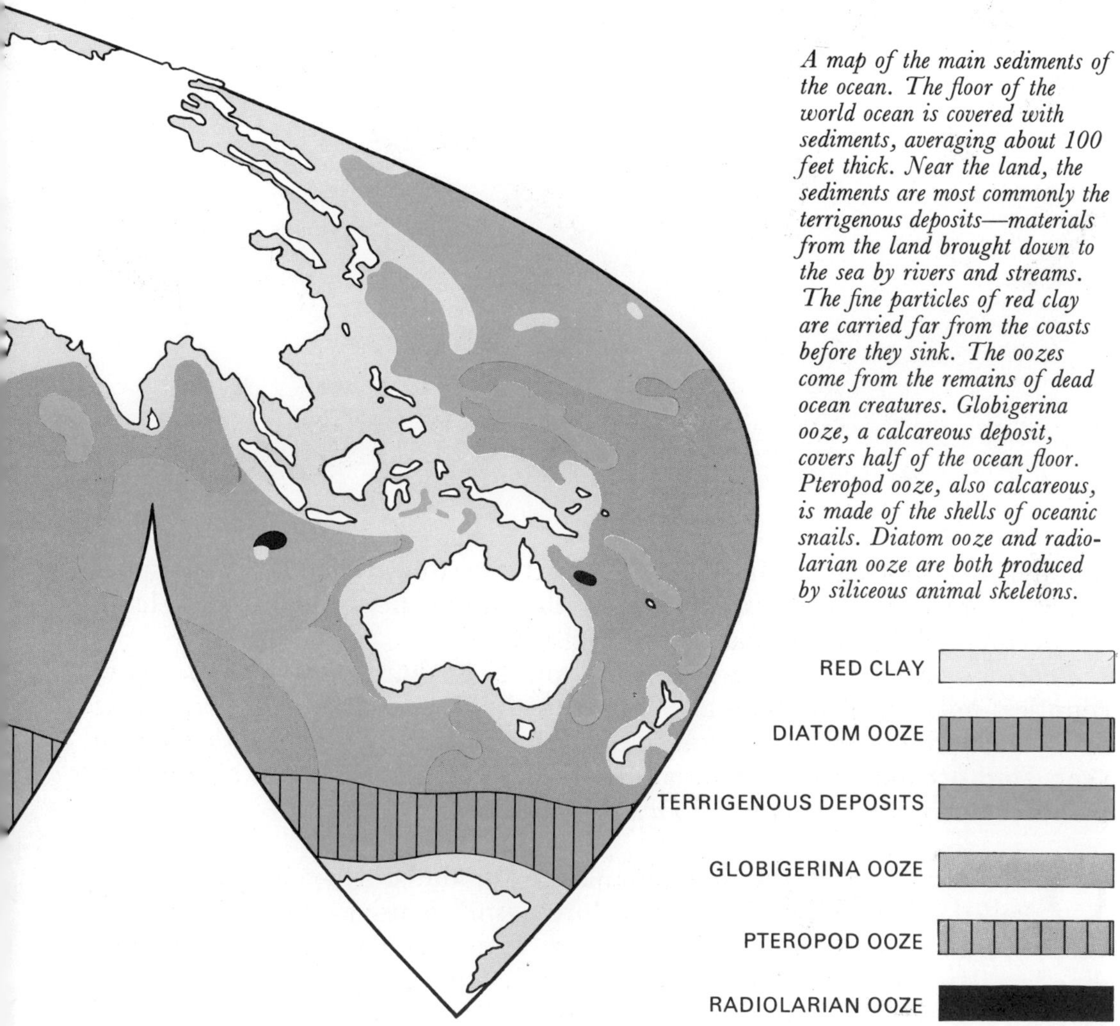

A map of the main sediments of the ocean. The floor of the world ocean is covered with sediments, averaging about 100 feet thick. Near the land, the sediments are most commonly the terrigenous deposits—materials from the land brought down to the sea by rivers and streams. The fine particles of red clay are carried far from the coasts before they sink. The oozes come from the remains of dead ocean creatures. Globigerina ooze, a calcareous deposit, covers half of the ocean floor. Pteropod ooze, also calcareous, is made of the shells of oceanic snails. Diatom ooze and radiolarian ooze are both produced by siliceous animal skeletons.

RED CLAY

DIATOM OOZE

TERRIGENOUS DEPOSITS

GLOBIGERINA OOZE

PTEROPOD OOZE

RADIOLARIAN OOZE

to handle—surely a strange by-product of the phytoplankton of past ages.

Diatoms flourish and reproduce themselves at an impressive rate (about once every 24 hours) wherever there is a good supply of nitrates and phosphates in the surface water. These nutrients vary enormously from place to place and from season to season. A typical springtime phosphate value for the North Sea is roughly 15 to 20 parts per million by weight, but in the rich South Atlantic waters around the island of South Georgia, the phosphate count can be over 100 parts per million. Often, however, the phosphate value is much lower, mainly because diatoms have already mopped up much of the phosphate in a tremendous burst of growth and reproduction.

Another substance that diatoms need is silica, for their skeletons. Oceanic seawater contains at most 400 parts of silica per million in surface waters. In very productive zones like the one just mentioned, it can happen that growth and reproduction of diatoms are prevented toward the end of

summer by a shortage of silica. One Antarctic diatom, called *Corethron*, starts in the summer robust and spiny, but by the end of the summer the skeleton has become spineless, and has very thin silica walls.

There are several thousand different species of diatom, each made up of countless millions of individuals. Together they make about three quarters of all the organic carbon that marine plants produce by photosynthesis each year. And oceanographers estimate that the total production of all the tiny plants of the ocean is between *15 and 20 thousand million tons!*

Other members of the phytoplankton are the single-celled organisms called *dinoflagellates*. The word "organism" is used quite deliberately, because although some of this group are plants, others are animals. The dinoflagellate which has the scientific name *Ceratium* is a plant, but unlike the diatoms it is able to swim by using its two *flagella* or whip-like organs. One of these it carries in a prominent groove that virtually encircles its body, while the other projects from a much smaller groove at right angles to the other one. It also has a light-sensitive spot, a sort of primitive eye, by which it can tell where the light is strongest, and swim toward it, thus keeping near the surface. Instead of a silica skeleton it has a tough outer covering of *cellulose* (as do the cells of higher plants), armed with spikes. This makes it difficult for enemies to swallow it. The two pictures of *Ceratium candelabrum* and *Ceratium tripos* opposite show what awkward mouthfuls they would be for a small animal. Some small animals actually avoid dinoflagellates but do not hesitate to engulf a spiky diatom, possibly because the diatom's spikes are too brittle to do any damage.

Dinoflagellates live in the surface waters of all the oceans, but they are most numerous in warm areas relatively poor in nutrients, where they may outnumber diatoms. However, they are not nearly such important primary producers as diatoms. And, because they have no mineral substance in their shells, they do not form a bottom deposit in the way that diatoms do.

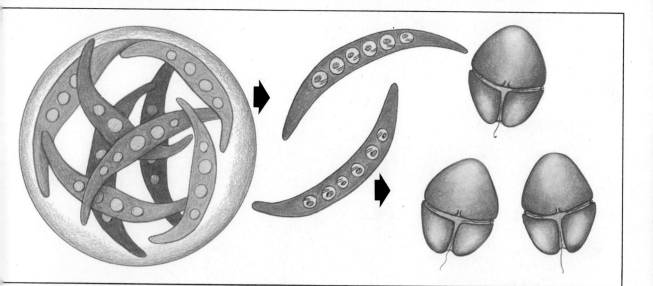

Above left: Zooplankton, here mostly copepods. They are the second stage in the food chain. Feeding on the phytoplankton—the primary producers—they are in turn eaten by the larger creatures of the world ocean.

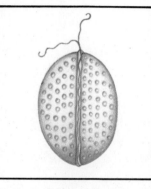

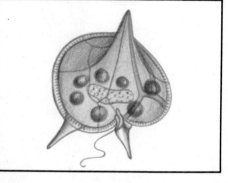

Top: A drawing showing the mode of reproduction of Gymnodinium, a dinoflagellate. It is a member of the zooplankton that forms a reddish coloration on oceanic water, the red tides that cause wide devastation. Gymnodinium releases into the water poisonous substances that are lethal to the other living creatures that come into contact with them.

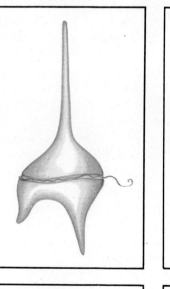

Right: Six other dinoflagellates. Top: A species of Exuviella *(left), and* Peridinium depressum *(right). Middle:* Ceratium Candelabrum *(left), and* Ceratium tripos *(right). Bottom:* Peridinium globulum *(left), and* Polykrikos schwarzi *(right).*

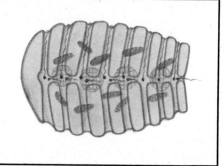

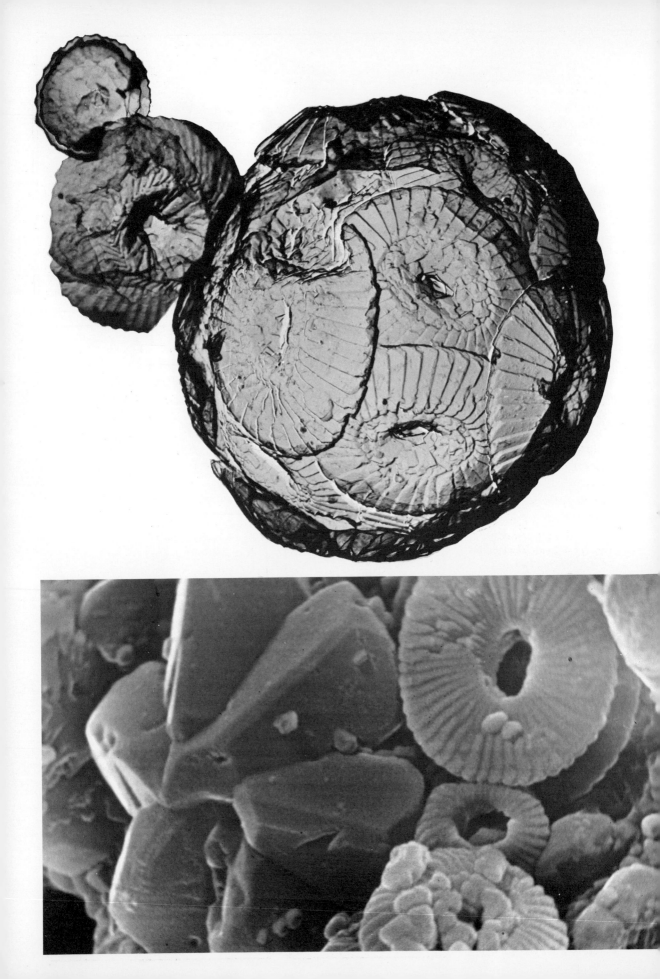

While most of the plant-like members of this group are harmless, there are certain dinoflagellates that spell disaster—such as those that go by the scientific names of *Gonyaulax* and *Gymnodinium*. In tropical and temperate waters these creatures can suddenly multiply so rapidly that the sea is colored red, and fishermen call this a "red tide." The huge accumulation of individuals, often amounting to well over three million per pint of surface water, is highly poisonous, and a severe outbreak of red tide results in the death of almost all other living things. Not only are *Gonyaulax* and *Gymnodinium* poisonous—they also release poisonous substances into the water. In shallow water, mussels absorb the poison from these dinoflagellates after having eaten them and store it in the liver. Any animal, be it bird or man, that eats such mussels may become seriously or even fatally ill from paralytic shellfish-poisoning. Luckily these outbreaks of red water are local and not very common. *Gonyaulax* plainly has a bad reputation, but it is only fair to add that quite recently it has been discovered that this flagellate contains a powerful antibiotic.

The dinoflagellates so far mentioned are clearly plants, because they contain chlorophyll and generate their own food: they are *autotrophic* (self-nourishing), like diatoms. But there are other members of the dinoflagellate family that have very much the same structure but live mainly on other plants and even on very small animals. Their way of life is called *heterotrophic* (feeding on others). One remarkable dinoflagellate, found mostly in temperate seas, is *Noctiluca* (meaning night-light). It is one of the largest of this group, being up to $\frac{1}{16}$ inch (the size of a pin's head) in diameter, and it is a voracious predator, feeding on diatoms and even on very small animals of the zooplankton. Its flagellum is modified for catching rather than swimming, and it glows brightly when the creature is disturbed. We do not know whether this light is intended to scare off predators or to attract zooplankton animals, but we do know that when huge Noctiluca swarms develop, usually in the late summer, every wave crest seems to be on fire. The wake of a ship through such waters glows so brightly that some people have claimed it is possible to read a newspaper over the stern on a dark night. However that may be, it is fascinating to lie at anchor in a small boat, and to see every link of the anchor chain and even the anchor itself clearly outlined by sparks, as the boat swings slowly on the tide. Biologists call this "living light" *bioluminescence*.

Noctiluca is among the biggest of single-celled animals. In fact there is a limit to the size that such creatures can reach. That limit is set by the amount of surface area through which they can absorb nourishment (and discharge wastes) compared with the *volume* of cell that has to be provided for. This *surface-volume ratio*, as it is called, becomes extremely important with increasing size. This is clear if we imagine a cell as a cube: double its length, and its surface area is multiplied by eight. So the bigger the cell the smaller is its surface area compared with its volume. This is why multi-celled

Two pictures of the fossil remains of minute one-celled plants, coccolithophores. In life they floated near the surface—as do coccolithophores today—but when they died they sank to become gradually embedded in the sediments of the ocean.

animals have large surface areas for absorption, such as lungs for gas exchange, or a lengthy gut for assimilating food.

We have seen how *Noctiluca* catches and digests its prey by means of its flagellum. There is also another device that single-celled animals can use to feed without actually chasing their prey. This is to send out threads of sticky protoplasm through holes in their armor, to form a kind of external gut that both traps and digests any diatoms or small animals that happen to come into contact with it. There are two quite separate groups of animals that use this method.

One of these groups is the Foraminifera, of which the commonest plank-tonic type is *Globigerina*. Most species live in tropical or subtropical waters, between the latitudes of 40°N and 40°S. Their shells are made of calcium carbonate, with numerous spines that help to keep them afloat. Another group, the Radiolaria, also feed by sending out threads of sticky protoplasm to catch and digest food. But they are like the diatoms in the way their elegant perforated shells and spines are made of silica.

With these two examples we come to the end of the minute creatures that succeed in feeding without actively hunting. Because of the surface-volume ratio, their size is roughly the limit for single-celled animals. Also single-celled animals larger than these would probably have great difficulty in

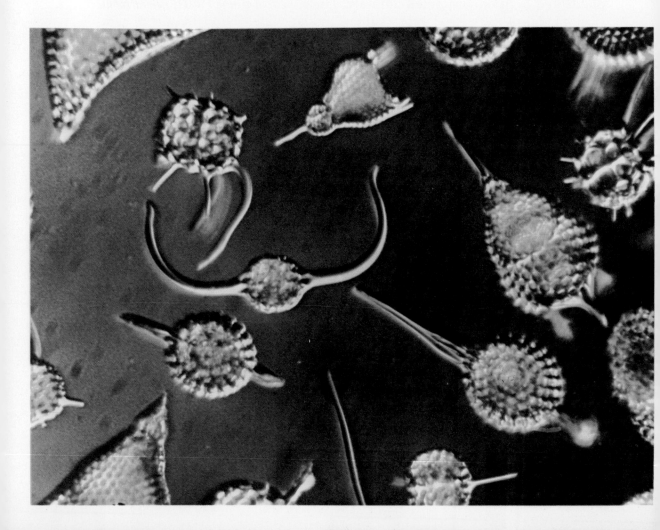

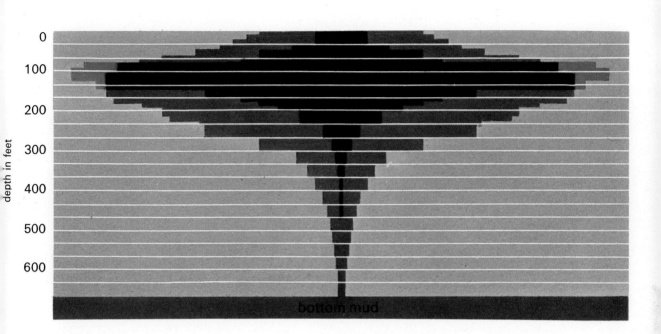

depth in feet

0
100
200
300
400
500
600

bottom mud

Above: Diatoms (green) and bacteria (orange) have a very similar vertical distribution in the sea, apparently because most bacteria are attached to the plant plankton. The overlap area is black. Bacteria are also found on the seabed, where they release nutrients from dead organic particles.

maintaining buoyancy. Most grazers on phytoplankton are many-celled, and much more complicated. They have developed by evolution to become mobile. They can therefore actively seek for food and the effort is worthwhile, even though this very effort of hunting and staying afloat consumes much extra energy.

Before leaving the minute creatures of the sea, it is worthwhile to note the extremely important part that bacteria play in the sea's economy. Bacteria are found in all seas and at all depths, but live mainly in the photic zone and on the seabeds of shallow coastal waters. They digest *detritus*, which we can think of as the rubbish of the sea. Detritus can be of two kinds. One kind consists of suspended particles of clay, mud, and other inorganic substances washed down by rivers, together with dust from deserts, volcanoes, atom bombs, and burned-out meteorites. The other kind of detritus is organic, consisting of dead and decaying particles that come from living organisms.

When we say that particles are decaying, we are really saying that they are being attacked and broken down by bacterial action. Because the photic zone is teeming with living things that must all eventually die, it is also a vast graveyard that supports a huge population of bacteria, varying from a few hundred to several thousands per milliliter of water. These marine bacteria live by breaking down plant and animal remains and releasing

Left: Radiolarians, single-celled animals of the plankton. Their skeletons are immensely complex, the simplest being latticework spheres and the most complicated a series of perforated spheres, the innermost one nested inside the others.

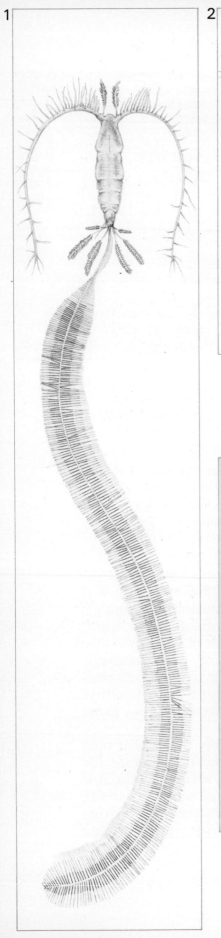

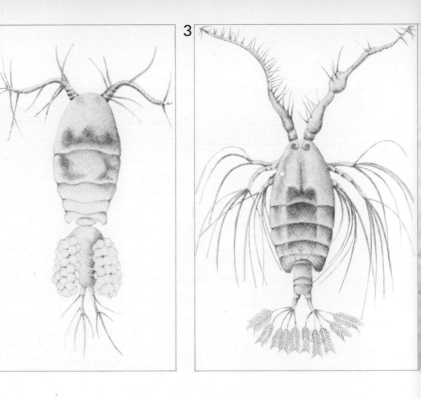

Above: Three copepods from the Mediterranean Sea: 1, Calocalanus plumulosus; 2, Oncaea mediterranea; 3, Pontellina plumata.

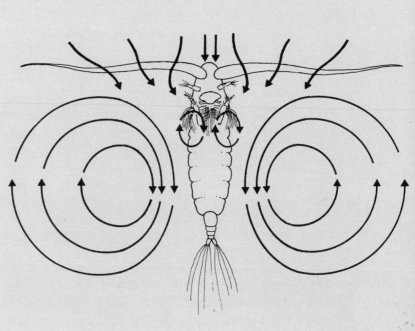

Above: The copepod Calanus finmarchicus, showing how its method of motion helps it to feed. The tiny currents it produces by moving forward also serve to swirl a stream of water into its filter net, which extracts its phytoplankton food.

inorganic salts into the sea, in the same way that soil bacteria break down organic matter into simple salts that can be absorbed by the next generation of land plants.

Phytoplankton can rightly be called the "pastures of the sea," because they are the primary source of food on which all other marine life ultimately depends. What are the links between the microscopic plants of the phytoplankton and large animals such as fishes and even whales? Clearly the first link must be the grazers—animals that feed directly on plants that are at most 1/25 inch long, and usually a great deal less. How much energy would such a grazer use if it relied on picking up one diatom here and another there? Would it spend more energy than it would gain?

The problem is best brought into focus by thinking first of a land equivalent of phytoplankton, and the terrestrial animals that live on it. The nearest land equivalent is probably a pasture that is constantly being grazed. If the pasture was properly managed it would consist of millions of blades of short grass together with other plants such as clover. Whatever its composition, the plants would be only a few inches tall and would be grazed as fast as they grow. Such a pasture would be of no use to an elephant, but it would be very useful to a small grazer, such as a rabbit. However, even a rabbit could not get enough to eat by nibbling off individual blades of grass and swallowing them. The effort would be more than the nourishment was worth. So the rabbit bites off, say 20 blades of grass in a single mouthful. But it can do this only if the blades are close enough together.

Now let us investigate how the grazers of the sea make a living. Unlike the blades of grass, which are fixed and rooted in one position, the plants of the phytoplankton are helplessly adrift in their watery environment. So, for that matter, is the grazer adrift. But it is by no means as helpless. The question is: are the diatoms or other planktonic plants sufficiently thick in the water for the grazer to get a rewarding mouthful, like the rabbit? The answer cannot be a simple yes or no. If phytoplankton is extremely concentrated, the grazer could well gulp down a sort of vegetable soup, but in fact it does not do this. For some unknown reason, it prefers to browse in areas where the diatoms are not too concentrated but nevertheless plentiful. This means that the "instant mouthful" is not available. How, then, does the grazer meet its needs? Let us find out by looking at some of the most important grazers in the world, present in every sea—the group of *crustaceans* (crab- and shrimp-like creatures) called copepods. One of the most abundant of the 4,500 known species of copepods is one called *Calanus finmarchicus*. In the fertile waters of the Northern Hemisphere it is one of the most important converters of plant matter into animal protein. It is quite a powerful swimmer as plankton animals go, yet it does not catch diatoms by swimming after them. Instead, it draws a stream of water and diatoms toward itself, and makes the water pass through extremely fine filters that strain out even the very smallest organisms. The drawing on page 62 shows the mechanism. An ingenious feature of this *filter-feeding* is that the current of water drawn in exactly balances the animals tendency to sink. This means that the animal need spend no more effort in keeping its position than that already needed to pass a stream of water through its strainers. The swimming power of this copepod's legs is reserved for large movements, such as avoiding predators or changing its depth in the water in response to light and darkness.

Calanus feeds by the "vacuum-cleaner method," a method also used by many other invertebrates (animals without backbones), including those that spend most of their lives permanently fixed in one place, such as mussels and oysters. *Calanus finmarchicus* is the biggest and most conspicuous copepod, but it is only one out of thousands of species. Taken together, the copepods are the most numerous grazers in the world. In the northwestern Pacific the average copepod population during summer is over 4,000 per cubic foot of water at the surface (where diatoms are most plentiful). Off the coast at Murmansk, in the Russian Arctic, a figure of 8,000 per cubic foot is not uncommon in summer. These creatures are also found in considerable numbers down to

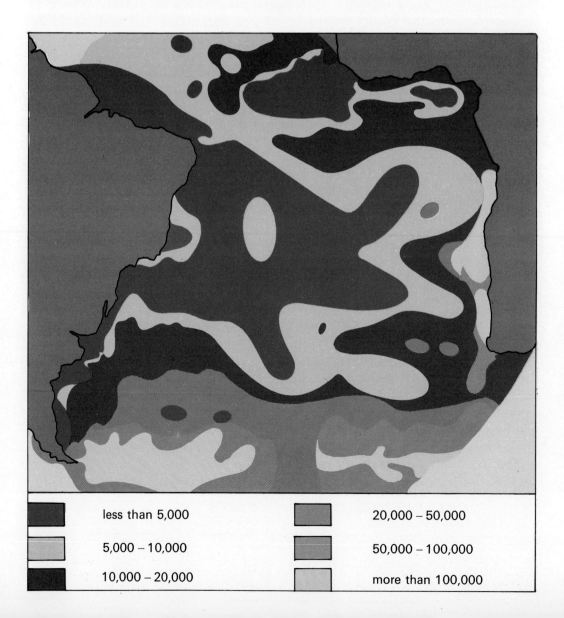

	less than 5,000		20,000 – 50,000
	5,000 – 10,000		50,000 – 100,000
	10,000 – 20,000		more than 100,000

16,500 feet, and in fertile Arctic waters such as the fishing grounds off Franz Josef Land there can be as many as 250 copepods per cubic foot at that depth.

These figures, however, are only examples. Copepods occur all over the world and in all climates, and in many areas they form the bulk of the total zooplankton population. For example, it is estimated that the average annual weight of zooplankton in the North Sea is 10 *million tons,* of which copepods account for about 7 million tons.

Another family of plankton animals is almost as important as the copepods, especially in the Antarctic. These are what can best be described as oceangoing prawns—the *euphausiids.* Copepods range in size from $\frac{1}{25}$ inch to $\frac{1}{5}$ inch, but the euphausiids are a good deal larger, ranging from $\frac{2}{5}$ inch to over 3 inches. Like copepods, they are found all over the world, but are most plentiful in cold waters.

In the Antarctic there are 10 known species of euphausiids. Largest of them all is *Euphausia superba.* This species grows to a length of over 3 inches

Left: A map showing the numbers of zoo- and phytoplankton that live in various areas of the South Atlantic, taking into account all plankton living down to a depth of 50 meters. The figures indicate the thousands of individuals in a liter of water.

Above: Krill, the shrimp-like creatures that form the food of whalebone whales in the polar seas. Each measures about 1¾ inches (including antennae). The movements of the huge whales are closely related to those of their tiny prey.

and looks very like a cooked shrimp, even in color. But whereas a shrimp turns pink when cooked, *Euphausia superba* is that color when alive, and this makes it easier for them to be seen when vast numbers are swimming close to the surface (not to mention deeper down). From the crow's nest of a whaling ship the author has seen long pink streaks perhaps as much as 100 feet wide and 1,500 feet long, where millions upon millions of *Euphausia* were swimming, all apparently in the same direction, and gorging themselves on a dense patch of diatoms.

The name given to large numbers of *Euphausia superba* is a Norwegian word, *krill*. The same word is also used in northern waters to describe the smaller euphausiids and large copepods on which northern whalebone whales feed. The krill of the Antarctic exist in such huge numbers that they dominate all other plankton in the waters south of the Antarctic Convergence, providing food for other plankton as well as for birds, seals, and fish. Above all they provide the staple diet of the great whalebone whales—among them the huge sulphur-bottom, or blue whale, probably the largest animal that has ever existed.

So between the phytoplankton and the great whalebone whales there is only one link, the krill that graze on the pastures of the sea. And we can use the diet of the big whales as evidence of the enormous amount of krill in the Antarctic. An adult blue whale, and its smaller relative the fin whale, each eat an average of four-fifths of a ton of krill per day during a summer season of about 150 days. This gives us a figure for the whole summer season of 120 tons per whale. (Whales eat very little during their southern-winter breeding season in the tropics). Before the mass slaughter of whales began, in the early 1900's, there must have been close on half a million large whales browsing steadily on the krill, so that they alone consumed about 60 million tons every year. But even then they can have eaten only a small fraction of the total stock of krill, for there are millions of penguins, other seabirds, fish, and seals also eating krill. Taking all these factors into account, it has been calculated that if we could work out an economic method of catching krill, we could take an annual harvest of 100 million tons—about twice the present yield of the world's fisheries.

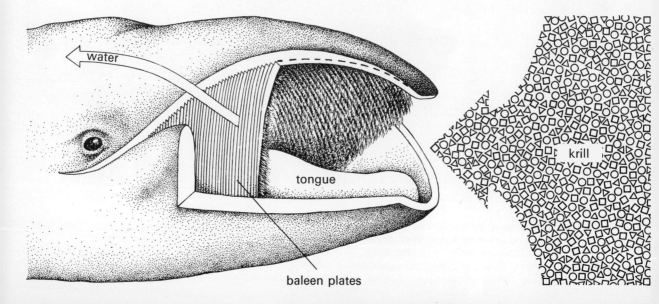

water

tongue

baleen plates

krill

Copepods and euphausiids are the main link in the food chain between plant matter and the rest of the life in the sea. But it would be wrong to imagine that they are the only one. Surface waters all over the world are, in varying degrees, the scene of greatest activity. And in these surface waters, there is a constant "eat or be eaten" battle going on.

It would be no exaggeration to say that nine-tenths of all plankton, whether plant or animal, is doomed to be eaten before it reaches old age and dies naturally. Copepods eat diatoms, large copepods eat smaller ones, and at the same time there are members of almost every other group of the invertebrates eating ceaselessly, only to be eaten in their turn by some other marine predator.

So great is the food supply in the surface waters that many animals that do not normally belong there make brief visits at certain periods of their life history. The full-time plankton of the deep oceans is called *holopelagic* (wholly open sea). But in shallow coastal waters there are also numerous young visitors from the seabed that temporarily feast on the surface bounty. During their limited stay, they grow quickly and gradually change their shape until they are ready to return to the seabed and settle down for life. These creatures are called *meropelagic* (partly open sea), and their stay among the surface plankton is usually quite brief—a matter of a few weeks. Although it often takes a microscope or a high powered hand lens to see these larvae, they swarm upward in their millions in order to get more food more quickly than they could at the bottom. Their way of life is a race against time. The parents lay eggs by the million, and when the larvae hatch they must start eating right away or perish. So they swim into the "marine cafeteria" where their only chance of survival is to grow quickly enough to avoid being eaten by other small predators. Of course, millions of them do get eaten, but those that survive grow—sometimes through a number of planktonic larval stages—until they are ready to sink back to the bottom and establish themselves. Incidentally, during the growing stages they may be carried far from their parents by currents and tides so that they eventually establish themselves on new areas of the seabed. In this way the tides and currents disperse many species over wide areas, just as winds disperse thistledown and thistles over the land.

Among the bottom-living animals that have meropelagic stages are a great many invertebrates, including many mollusks, starfish, sea urchins, sea cucumbers and bristle-worms. Certain *vertebrates* (backboned animals) that inhabit deep waters as adults also go through a meropelagic stage when young. For instance, the bottom-dwelling flatfish called flounders lay buoyant eggs that float up and hatch in the surface layers. The young fish gradually change their shape. Their bodies end up flattened because their skulls gradually become twisted over on to the left side, and then they can begin their adult life of feeding on the bottom.

One interesting feature of these meropelagic forms is that at first sight it is almost impossible for anyone but an expert to guess what form the adult

Instead of teeth, the baleen whale has long, thin fringed strips of bone called baleen plates. When the whale gulps a mouthful of water, its tongue lifts and forces the water out through the sieve-like plates, straining out the krill, which it eats.

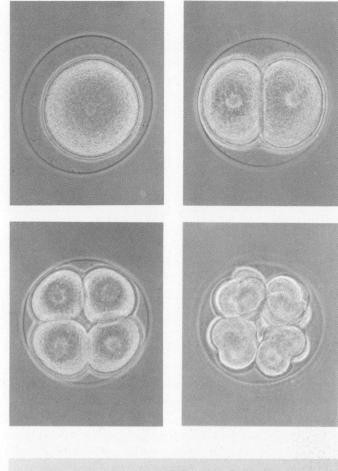

Right: A by-the-wind-sailor,
Velella, *being attacked by a
snail,* Janthina. *The drifting
ocean life of the surface is
as fiercely predatory as the
other areas of the seas.*
Left: The development of an
egg of the sea-urchin
Paracentrotus lividus. *Like
so many marine animals, the
sea urchin passes through a
planktonic larval stage before
metamorphosing into the adult
form, shown below. The sea
urchin is free-living all its
life, but in the larval stages
drifts with the currents, as a
member of the plankton, on
which it feeds.*
Below right: By-the-wind-
sailor drifting on the surface.
It is closely related to the
Portuguese man-of-war, but its
sting is much less potent.
Below center right: The
beautiful but deadly Portuguese
man-of-war. It stings with a
nerve poison with some effects
similar to cobra venom.
Below far right: The purple sea
snail Janthina, *which hangs
on a mucous raft of air bubbles.*

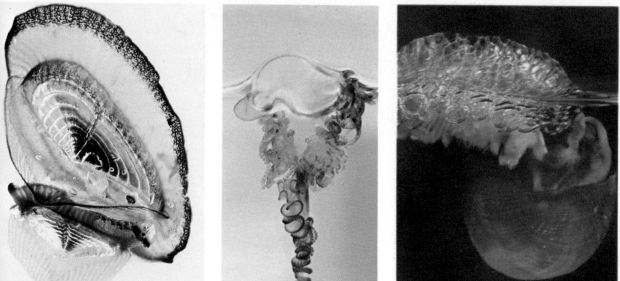

will take. For example, the delicate larval sea-cucumber in the picture at the beginning of this chapter is destined to change into a slug-like adult.

When talking of life in the surface waters, where most animals of the ocean live, oceanographers are using the word "surface" in a relative way, to distinguish between the well-lighted upper waters and the depths of the sea. But there are some animals that not only live right at the surface itself, but also show part of their bodies above the water. Sometimes biologists place these animals in a separate group under the name *pleuston* (from the Greek *pleo*—I sail).

The best known of these "sailors" is the Portuguese man-of-war (*Physalia physalis*). It looks like a single animal with an inflated bag protruding above the water, a thick hull just below the water, and a number of long tentacles trailing several feet below the surface. In fact it is a *colonial* animal, in which different parts of the colony perform different jobs. *Physalia* belongs to a group called the *coelenterates*, and these are among the simplest forms of multicelled life, consisting of little more than a surface layer and a hollow inside where food is digested. Jellyfish are coelenterates too, but whereas jellyfish move through the water by slow pulsations of their umbrella-shaped bodies (a sort of primitive jet-propulsion), *Physalia* sails through the water, trailing the long tentacles that act both as a keel and as food catchers. These tentacles carry hundreds of extremely powerful stinging cells. Any plankton animal, or even a small fish, that bumps into them is stung to death, and then hauled up by the tentacles into the digesting members of the colony which hang below the main hull.

Physalia normally lives in the calm equatorial waters, and it sails at such an angle that it remains in those warm regions. But because the currents and the prevailing light winds are the exact opposite of each other on each side of the equator the northern and southern varieties of *Physalia* are mirror images of each other. Sometimes, individuals belonging to the northern variety get blown off course into the Gulf Stream, and if the summer is hot enough they survive to reach the English Channel. At such times, warnings are broadcast to bathers, because the sting of *Physalia* can make an adult extremely ill, and has even been known to kill young children.

Physalia might seem to be free from attack from other marine animals because of its powerful stinging apparatus. But it has at least one enemy—a surface-living snail, *Janthina*. This inch-long ocean-going snail drifts on the surface supported by inflated bladders of *mucus* (much the same as the slimy material that a land snail leaves on dry ground). *Janthina* appears to come to no harm from *Physalia's* stings, and it is on record that two of these snails devoured a *Physalia* colony in 24 hours.

Another animal that is not harmed by the stings (or maybe avoids them) is the little man-of-war fish. This fish lives among *Physalia's* tentacles, and it is camouflaged perfectly, because its body is a silvery color striped with the same shade of blue as *Physalia's* tentacles. The man-of-war fish seems to have become adapted to living safely within the poisonous forest of *Physalia's* tentacles, which are a highly effective deterrent to the little fish's predators.

A close relative of the Portuguese man-of-war is the smaller by-the-wind-sailor (*Velella velella*), which also has two mirror-image varieties living on opposite sides of the equator. Like *Physalia*, it dwells mainly in warm tropical waters, and is only occasionally blown onto the European and North American Atlantic Coasts. It, too, is attacked by the predatory snail *Janthina*.

The Sargasso Sea is a vast whirlpool in the southern part of the North Atlantic. Its waters spin slowly but endlessly, rotated by the south-westerly stream of the North Equatorial Current and the northeasterly thrust of the Gulf Stream. Its center is roughly to the east of Puerto Rico, and its area varies with the variations in the Currents between about $2\frac{1}{2}$ and 4 million square miles. It is a warm, high-salinity zone where there is very little mixing between the surface water and the more nutrient-rich water below. The depth of the sea in this region is at least 9,840 feet.

Floating and trapped indefinitely in this whirlpool there are vast quantities of a seaweed that has small bladders filled with gas for buoyancy, not unlike the bladder-wrack of northern shores, which normally lives firmly attached to rocks, but is occasionally found out at sea after severe storms. This seaweed *Sargassum* (after which the area is named), is torn off the coasts of the West Indies, but unlike the northern bladder-wrack it survives for a very long time after it gets into the giant whirlpool of the Sargasso Sea. It even goes on growing, but it cannot reproduce once it has been separated from its rocky foothold. Nevertheless, some marine biologists believe that part of the estimated 4 to 10 million tons of floating weed found in this area was alive when Columbus drifted through the Sargasso Sea in 1492. Incidentally, the tonnage of floating weed is not so impressive as it might seem. It works out at only $3\frac{1}{2}$ tons per square mile, and there is no question of its being dense enough to trap sailing ships—as a popular legend of bygone days suggested. That legend was born in the days when sailing ships were often becalmed for days, and even weeks, in the Sargasso Sea. The crews were doubtless bored, and also frightened of running out of drinking water. And as they saw this weed, day after monotonous day—mostly as individual plants, but occasionally bunched together in larger clumps—the story gradually grew up that the Sargasso was a graveyard for ships.

The existence of this floating weed is unusual enough, but what is most interesting is that it harbors a mixed population of small animals that we can call plankton (because they are drifting), but that must originally have come from the shallow waters of the West Indies along with the weed.

We can be reasonably sure that the animals that either cling to the weed or shelter in its fronds were originally shallow-water Caribbean animals, because they are not found elsewhere in the deep sea. Among the 50-odd species that take advantage of this drifting home are bristle-worms, snails, shrimps, and crabs. If any of these were to leave the shelter of the weed

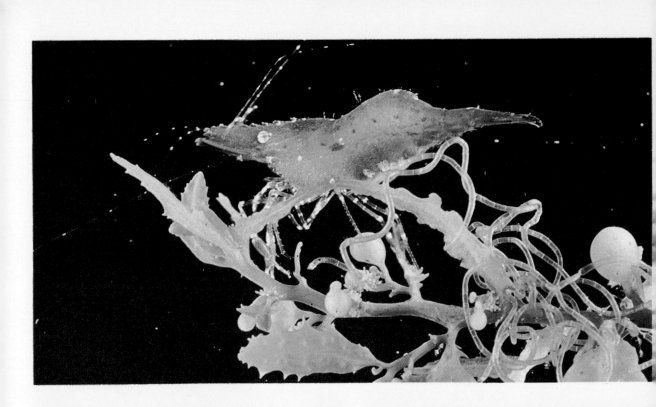

*Small animals of the Sargasso use the seaweed
for support and camouflage, taking on its colors.
Top: A prawn moving through the Sargassum weed.
Above: A camouflaged frog-fish (Pterophryne).
Left: A fish and an anemone from the Sargasso.*

they would either be snapped up by predators, or sink helplessly to the bottom. There are also a number of small fish, heavily camouflaged so that it is difficult to distinguish them from the weed itself.

Most of the animals found among Sargasso weed are dwarfed versions of the animals that inhabit the fixed *Sargassum* of the West Indies coasts. This fact points to two conclusions. One is that the food potential of the Sargasso Sea is very low, and certainly much lower than that of West Indian coastal waters, so that small size has a great survival advantage. The other is that these species have been drifting round and round with the weed, for perhaps many thousands of years, and over many generations have gradually adapted themselves to the conditions of the Sargasso Sea.

The Sargasso Sea can claim at least one other distinction. All the eels that inhabit European and Eastern American rivers and ponds are believed to start their lives 6,500 feet down in its center. It is to this point that the adult eels swim on their last long journey of well over 1,800 miles to lay their eggs, and then die. And it is from this point that the larval eels drift in the ocean currents until they reach the coasts of their future freshwater homes.

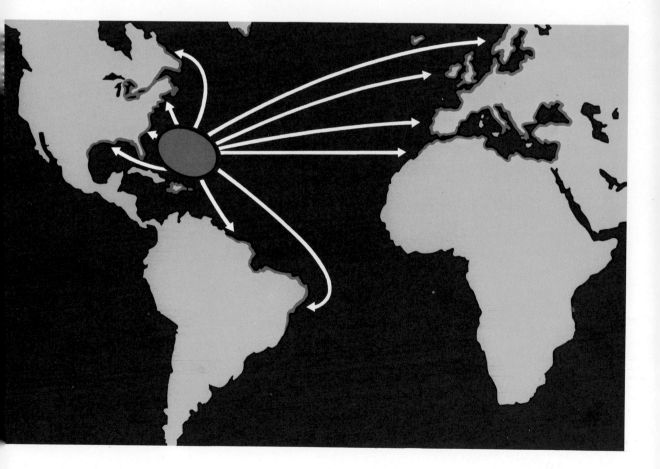

Above: The migration of young eels from their breeding grounds in the Sargasso Sea to coastal areas (red) where they live as adults. Although guesses had been made for centuries about the origins of eels, only in 1922 was the problem solved.

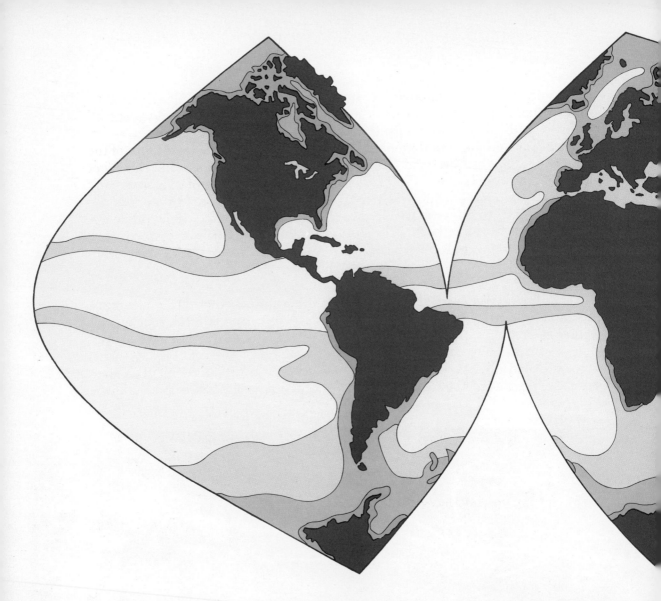

The adult European eels live in the rivers and streams of Europe and North Africa, while the American eels spend much of their adult lives in the fresh waters of the East Coast of North America. They are identical except for the number of their vertebrae, 110–119 in the European eel, and 103–111 in the American eel. The males move down to the sea when they are four to eight years old, but the females remain in the rivers until they are seven to twelve years old.

Before undergoing their long migration across the Atlantic the mature eels stop feeding; their heads become pointed and they change color from yellowish-gray to greenish-brown color on the upper surface and silvery-white on the belly. Also the eyes and side fins enlarge.

Although mature eels of the American variety have been observed migrating toward their Sargasso breeding-grounds, no European eels have been caught in the open ocean. Some biologists have suggested that the European and American eels belong to the same species and all the so-called "Euro-

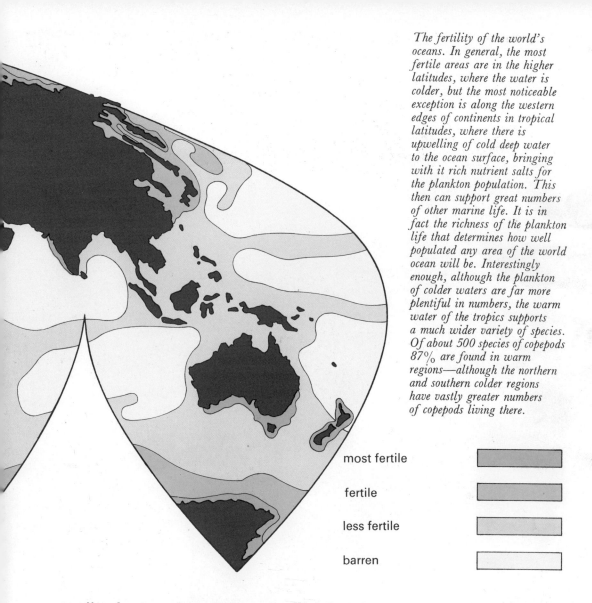

The fertility of the world's oceans. In general, the most fertile areas are in the higher latitudes, where the water is colder, but the most noticeable exception is along the western edges of continents in tropical latitudes, where there is upwelling of cold deep water to the ocean surface, bringing with it rich nutrient salts for the plankton population. This then can support great numbers of other marine life. It is in fact the richness of the plankton life that determines how well populated any area of the world ocean will be. Interestingly enough, although the plankton of colder waters are far more plentiful in numbers, the warm water of the tropics supports a much wider variety of species. Of about 500 species of copepods 87% are found in warm regions—although the northern and southern colder regions have vastly greater numbers of copepods living there.

most fertile

fertile

less fertile

barren

pean" eels grow from larvae that hatched from eggs laid exclusively by American eels, and that the adult European eels never reach the Sargasso Sea, but perish in the Atlantic Ocean. They also suggested that the difference in the number of vertebrae arises from different environmental conditions experienced by the larvae in their early stages.

The eggs laid in the Sargasso Sea hatch into leaf-like larvae that were formerly thought to be tiny fish of a completely different species and were given the name *Leptocephali*, which is still used by marine biologists. These tiny young eels drift with the currents of the Gulf Stream to the shores of Europe and eastern North America. The European larvae take three years to complete their 3,000 mile long journey, while the American larvae reach their much nearer goal in only a year. When they reach the coast they change into young eels, or elvers, also known as "glass eels" because of their almost transparent appearance. The elvers ascend the rivers and streams, where they grow to maturity.

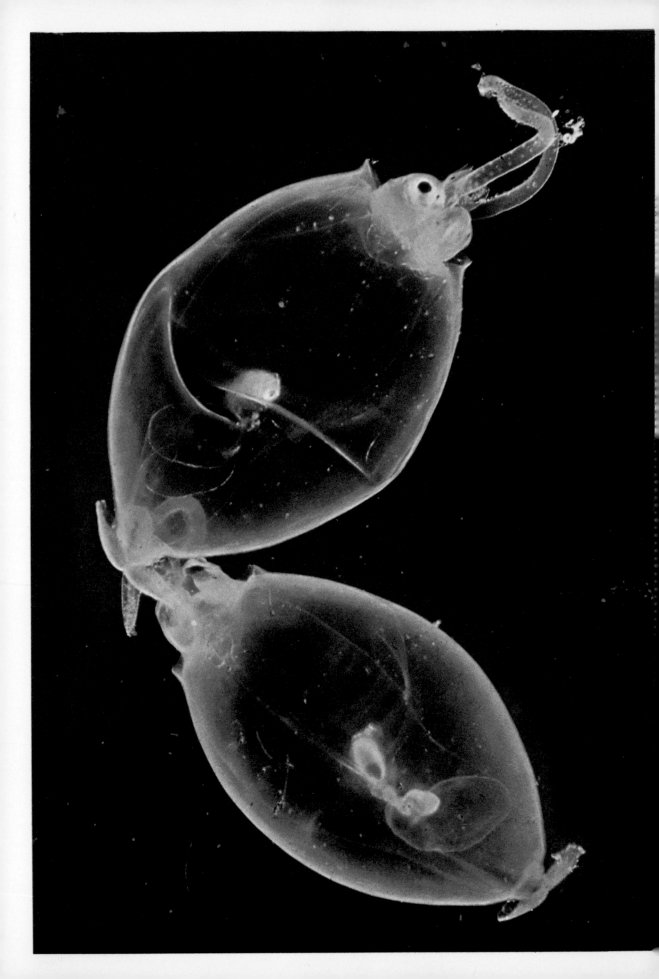

4

The Puzzle of Vertical Migration

We have already seen how certain marine animals spend part of their larval life in the upper, food-producing layer of the sea before settling down to adult life on or near the bottom. These animals that go through a drifting larval stage in fact make a vertical migration, although they do so only once in a lifetime. Theirs is, however, by no means the only vertical movement to be seen among sea creatures. There is also a 24-hour cycle that can be observed in many ocean animals, which marine biologists call *diurnal vertical migration*.

Broadly speaking—but this is a generalization, and should be treated with caution—the grazing animals spend the daytime hours below the surface of the sea, so that in bright daylight the surface waters, where the phytoplankton is most plentiful, contain very few grazers. As the sun sets, the grazers and other animals that prey on them move up into the photic zone, where they stay throughout the night. Then, as dawn approaches, they begin their migration down again, until by broad daylight the photic zone is again almost empty of animals. Intensive plankton surveys around the island of South Georgia in the South Atlantic, and expeditions such as that in 1910 of the Norwegian research ship *Michael Sars* in the North Atlantic, have given marine biologists much information about the daily journeys of some zooplankton species.

In the early days of marine biology, when scientists started to use plankton nets, many of them noticed that their catches of plankton in the surface waters were far greater at night than those they made during the daytime. Some marine biologists tried to explain this by saying that, during the hours of daylight, the plankton creatures could see the approaching net, and were able to escape it. At night, on the other hand, they were unable to detect the net until it had closed round them. This is true for some large and powerful young fish, but it simply does not hold for the majority of plankton creatures.

By taking samples with plankton-nets towed at different depths at the same time marine biologists demonstrated not only that there are *more* plankton creatures in the upper layers at night, but also that there are *fewer*

Two transparent larvae of the squid Cranchia. *Nightly they make a vertical migration to the surface waters, where they feed during the dark hours. As the sun rises, they sink back to the mid-waters, between 300 and 1,500 feet deep.*

of them in the lower layers at that time. When the towing of the nets was completed, the end of each net was closed by a special mechanism before being hauled up. This was to make sure that all the animals in each net had been caught at the depth at which that particular net had been towed, and had not been caught on the way up as the net was being hauled in. Depth gauges were also used to record the exact depth of each net.

Other researchers demonstrated vertical migration in the laboratory. In 1954 two British marine biologists, Alister (now Sir Alister) Hardy and Dr. Richard Bainbridge invented an ingenious device known as the "plankton wheel." This allowed them to watch plankton animals making their vertical migrations and also enabled them to determine the speeds at which they were swimming. Also, a record could be made of every upward and downward movement of the tiny travelers, and the light intensity, which affects vertical migration, could be altered. The swimming speeds of some plankton creatures are quite astonishing. For instance, the little copepod called *Calanus*, no bigger than a grain of rice, can climb almost 50 feet in an hour, while certain euphausiids are even faster, traveling at over 300 feet an hour! In the course of these experiments, it was discovered that the plankton animals swim downward considerably faster than they are able to climb upward. This discrepancy is accounted for by the fact that all plankton animals (except those that float at the surface) are slightly heavier than water.

The plankton wheel and other devices for studying the migration of the

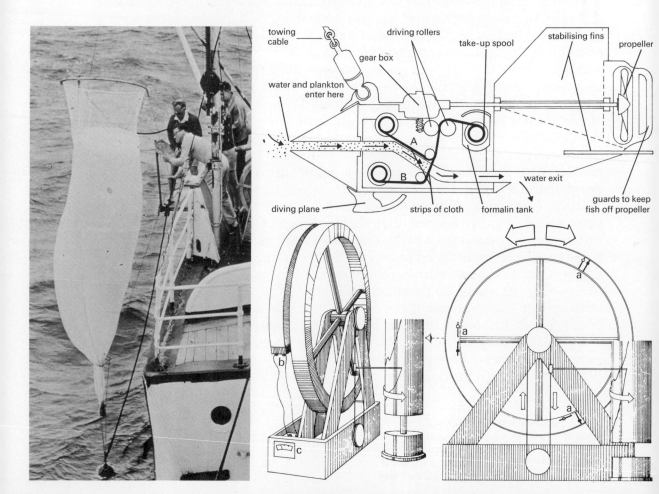

plankton are illustrated below. All the results of studies of vertical migration have shown the same general pattern: grazing animals of the plankton migrate to the surface waters when night falls, after spending the day at deeper levels. For instance, a copepod that has the scientific name of *Clausocalanus laticeps* is found at depths of 200 to 500 feet during the day, and in great numbers in the top 65 feet of water around midnight. Other copepods migrate from a depth of 650 feet and back again.

Although most copepods travel up and down to some extent, there are several exceptions. A number of species live mainly between 150 and 500 feet, and make no vertical migration at all. Others move upward at night, but not as far as the surface.

Some grazing animals of the plankton never stray far from the surface. The krill (*Euphausia superba*), which we met before as the food of the great whalebone whales in the Antarctic, are a good example. They do make some vertical migration, but only between 200 feet and the surface.

So *Euphausia superba* spends more time in the phytoplankton zone than do most other animals of the zooplankton—at least when it is adult. But this is not the case with all the creatures of the zooplankton. The close relatives of the krill, *Euphausia frigida* and *Euphausia triacantha*, also from the Antarctic Ocean, remain well below 600 feet during the hours of daylight and then, in an hour or two, rise to between 330 feet and the surface at sunset. At sunrise next morning, they return just as quickly to the deep water.

Far left: A plankton net being lowered over the side of a research vessel. By lowering nets to different depths it is possible to study the vertical distribution of the plankton.

Below left: The "plankton wheel." The wheel is a transparent tube of perspex, which can be filled with water and plankton, and turned so that the animal is halfway up one side, and its movement up or down can be observed. As it moves, the wheel can be turned by hand so that the animal is kept in the same relationship to an observer, while actually swimming in an endless tube. Little doors, marked a, are used to keep the water moving exactly with the tube, rather than lagging behind it. b is a photo-electric cell, and c a meter—they record the light intensity at a point just beside where the animal is swimming. A drum-recorder, turned by an electric clock, keeps a record of the movement of the wheel, and therefore the speed at which the animal is moving.

Right: A "fish," containing a sonic depth-finder, can be towed behind a ship and used to record the depth and size of plankton colonies.

Above left: The Hardy Plankton Recorder, a continuous device powered by its own propeller, designed to be towed under water by merchant ships not otherwise engaged in research. The plankton is collected on two silk gauze rolls, marked A and B, which are then folded together with the plankton between them, and wound onto a take-up spool. There the plankton sample is preserved in a tank of formalin.

The vertical migrations undertaken by different plankton animals are shown in the diagram below. It shows that some creatures live on the seabed itself in the daytime, moving upward toward the surface at nightfall. They may not be able to reach the surface waters before day breaks and the light drives them downward again.

Diatoms, like other members of the phytoplankton, normally remain in the photic zone, and do not migrate vertically like those animals of the zooplankton described above. Indeed, as we noted earlier, the spikes that most diatoms possess give them a large surface area for buoyancy and also a better "grip" on the water, so that they can *stay* near the sea's surface in the photic zone. There is, however, one diatom that behaves very differently from the rest. It is known to science as *Coscinodiscus bouvet*, and has been studied in the Antarctic waters off the island of South Georgia.

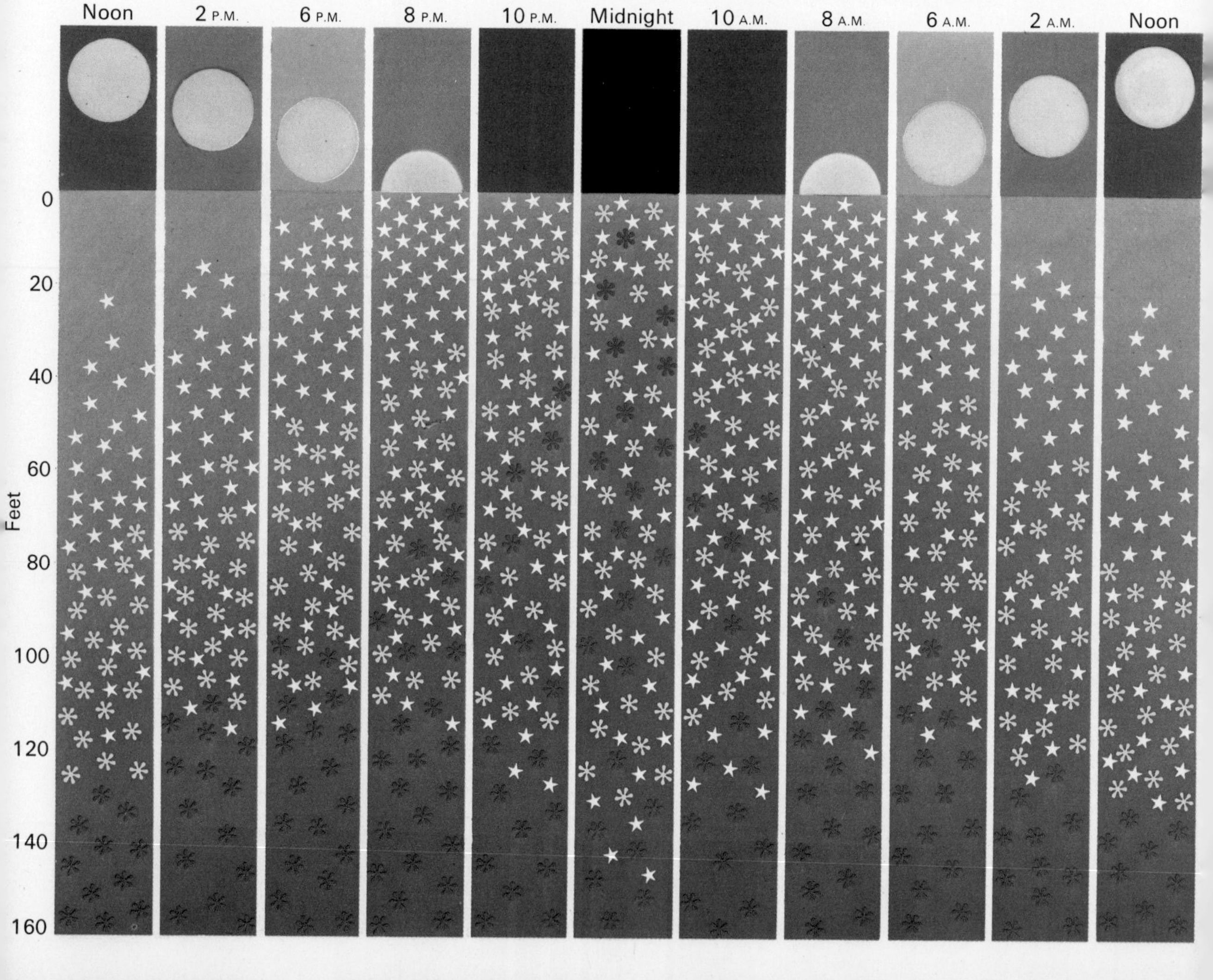

Coscinodiscus bouvet is a pillbox-shaped diatom studded with tiny holes but without spikes. It is far from plentiful among the Antarctic phytoplankton, and accounts for less than one sixteen-thousandth of the total number of diatoms of different species caught in plankton nets. This exceptional diatom has the unique habit of migrating daily in exactly the *opposite* direction to that of the zooplankton. During daylight it remains near the light, where it carries out photosynthesis, and at night it descends to between 320 and 820 feet, just when the grazers are coming up to feed. This is shown in the diagram on page 87.

One possibility is that toward the end of a daytime's photosynthesis, some of the sugar it produces is converted into starch, which is insoluble in water and also more dense than sugar. This would result in a loss of buoyancy, and the organism would sink. Then, during the night, the diatom may make comparatively light-weight fat from the starch, so that toward morning its buoyancy is increased to such an extent that it rises.

The other possibility is that *Coscinodiscus* produces bubbles of carbon dioxide gas by respiration during the night, and that by morning these bubbles give it enough buoyancy to rise to the surface. There, by day, the carbon dioxide would be replaced by bubbles of oxygen during the process of photosynthesis. Then, as the light fades and photosynthesis gradually stops, the supply of fresh oxygen would diminish, because oxygen is constantly being used up in respiration causing this unusual diatom to sink again. Whatever the mechanism of the process may be, the spectacle of *Coscinodiscus bouvet* "dodging the rush-hour" is an amusing and remarkable example of an adaptation, brought about by the long process of evolution, to the struggle of life.

Right: A photomicrograph of the diatom Coscinodiscus grani, *a close relative of the curious Antarctic species* Coscinodiscus bouvet, *which apparently makes a daily vertical migration exactly opposite to that made by most of the zooplankton.*

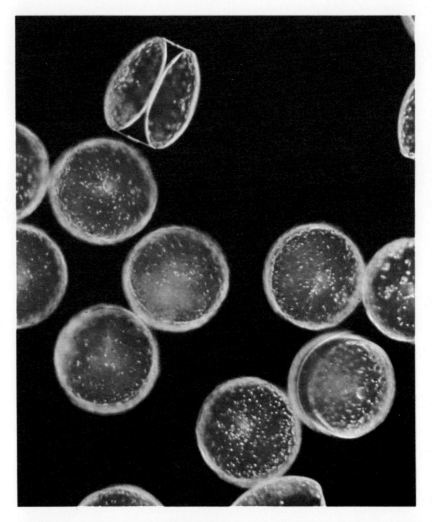

Left: In the general movement of vertical migration, each type of tiny plankton animal has its own pattern. Here white stars represent the copepod Calanus finmarchicus, *the principal food of the herring; yellow asterisks stand for a jellyfish,* Cosmetira pilosella; *and the red asterisks show the shrimp-like* Leptomysis gracilis, *which lives on the bottom during the daytime.*

Left: A sonic depth finding machine showing a recorded echo trace. The line shows the depth from which the echo returned, and therefore produces a "drawing" of the profile of the seabed.

Right: A sonic depth finder in the ship's hull emits a pulse of sound that travels downward through the water until the sound waves hit a surface that reflects them. In this case it is the ocean floor that is hit. Far right: Large shoals of fish, particularly those such as the lantern fish with swimbladders, reflect a great deal of the sound and are called the deep scattering layers, capable of returning very strong echoes.

We have so far been concentrating on the vertical migrations of animals that depend—either as grazers or as predators—on the rich population of the photic zone. All these migrations have a great deal to do with the change in light intensity from daylight to darkness. But the dimmer waters below the photic zone, down to about 3,300 feet, also support a dense population of animal life, part of which is plankton and part *nekton*. This word (from the Greek *nektos*, swimming) is used to describe all animals that can swim purposefully and strongly enough to make themselves less dependent on ocean currents than the plankton.

The zone between 650 and 3,300 feet is called the *twilight zone*. Human beings in a bathyscaphe at 3,300 feet would say that it was pitch-dark at this depth, except perhaps in very clear water with the sun directly overhead, but photoelectric detectors can register a very small amount of blue and blue-green light. Even this tiny amount of light may be enough to trigger off vertical migrations of both plankton and nekton.

The plankton in this zone consists largely of predators (though some are omnivorous—that is, they eat anything). For plankton animals, they are fairly powerful swimmers. For example, some euphausiids and deep-sea prawns make considerable nightly migrations toward the food-producing zone. The strongly swimming fish and squids come right up to the surface during the hours of darkness.

Information on the migration of twilight zone animals was gathered only gradually until World War II. Then new evidence suddenly came from an unexpected direction. Various *sonic depth finders* (or echo sounders) of various kinds had been in use since the 1920's, for studying the shape of the ocean

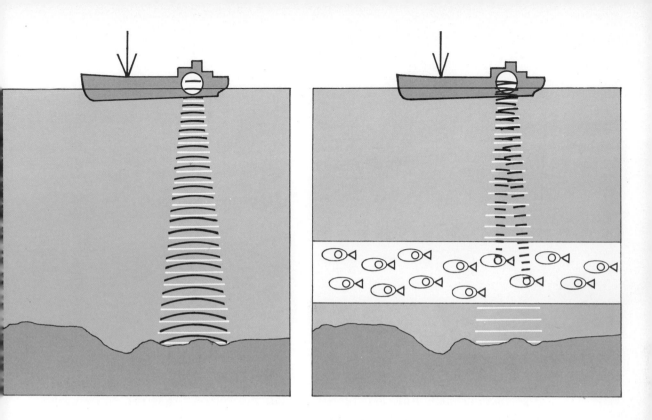

bottom. By measuring the time it takes between sending out a pulse of sound and receiving an echo from the seabed, it is possible to calculate the distance to the bottom at any given point. By the mid-1930's *ultrasonic* depth finders (those producing sound too high for humans to hear) were providing more accurate information, and were also being used by fishing fleets for detecting shoals of cod and herring in shallow inshore waters.

Ocean-going ships using similar equipment, however, began to report that shallow soundings were being recorded where the charts showed deep water. With the onset of World War II, physicists of the U.S. Navy started to investigate these mysterious "false echoes," as they were called, as part of their development of submarine-hunting techniques. Conducting experiments in the deep waters of the Atlantic and the Pacific, they found that false echo readings were often obtained at depths from 650 to 3,300 feet. Oceanographers were baffled, and thought that the echoes must have been caused by physical variations in the water, such as sudden changes of density. But all the observations had one thing in common: echoes came from 650 to 3,300 feet during the day, and from levels much closer to the surface at night. Finally an American marine biologist, Martin Johnson, was able to link the occurrence of these false echoes with what was already known about the vertical migration of plankton. And scientists now generally accept that the readings are obtained from echoes returning from various marine organisms that change their depth between daylight and darkness.

The groupings of these organisms that cause the false echoes are known as *deep scattering layers*. The fact that strong echoes are received suggests that the animals making up the deep scattering layers must be present in great

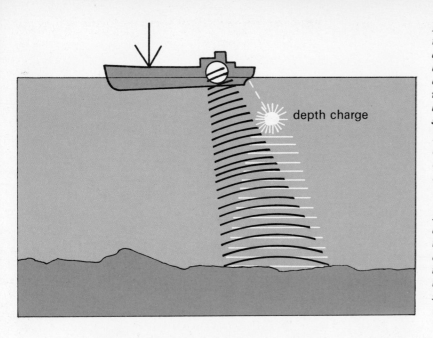

depth charge

numbers, but sampling has shown that the concentration of planktonic animals is far smaller at depths between 650 and 3,300 feet than it is near the surface. So it is now accepted as much more likely that the echoes are caused by certain structural features of the organism. The external skeletons of some crustaceans, for instance, could probably act as a very effective sound reflectors.

Again, many planktonic creatures contain gas-filled spaces, and large groups of them might resound like drums when the pulse of sound from the ultrasonic depth finder reaches them, sending back strong echoes. The gas-filled swim-bladders of certain fish also reflect sound puless. These fish have been traced by echo-sounding, and are caught by mid-water trawls at 820–720 feet by day and captured in the surface waters at night. Sometimes two or three separate layers can be detected, one beneath another. This could well be caused by the groupings of different species at different depths.

It is evident, then, that ultrasonic depth finders can provide clear evidence about the movements of certain fish and planktonic organisms. But we are still not sure what other animals may mass together in large enough numbers to give an echo. Small squids, for example, may well be the cause of echoes, but they are extremely difficult animals to count. Their jet propulsion systems make them extremely agile, and the numbers caught when nets are lowered are probably only a small fraction of those that escape.

Certain jellyfish are known to cause echoes but their fragile bodies are destroyed by the nets.

Marine biologists are now getting more accurate results than those from depth finders by setting off small-scale underwater explosions. Ultrasonic depth finders send out an impulse on only a single frequency, and such frequencies do not necessarily bounce an echo off animals of different sizes. Underwater explosions, however, produce a broad band of frequencies, and by using a sensitive receiver that is not tuned to any one frequency, multiple

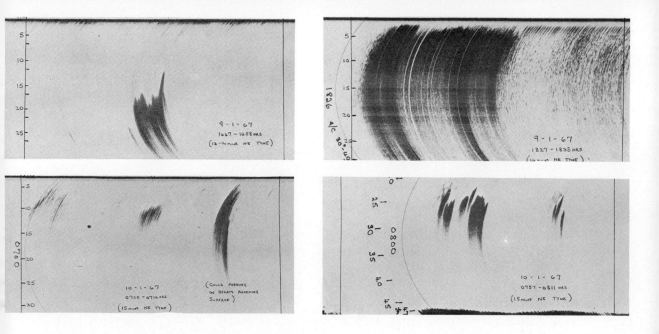

deep-scattering layers have been detected, and more detailed information on vertical migrations is being collected.

We are still far from knowing *why* so many species of zooplankton migrate vertically every day. Such long journeys consume a good deal of energy, and it is hardly likely that the habit would have been evolved by almost every major group of plankton animals, as well as by many fish, unless it had real survival value. If we look at vertical migration with this in mind perhaps we can find some explanation.

Both laboratory tests and research at sea with plankton nets have shown that there is little doubt that the movements of migratory zooplankton are controlled chiefly by the amount of light to which they are exposed. The animals are driven down by daylight and can move up into the photic zone only in the absence of light between sunset and dawn.

Biologists have also found that during the daylight hours different species consistently occupy different depths. It seems very likely that this "layering" is also directly related to light, because during the night the different species cease to occupy their distinct layers and mix freely in the surface waters, where they feed on phytoplankton and on each other.

The depth to which plankton animals sink does not vary only according to species. In summer, when there are more hours of daylight, and when the light is more intense, they are generally found at lower depths than during the winter. In the high latitudes of the polar regions, during the seasons when it is either permanently dark or permanently bright daylight, there are almost no migrations. The depths occupied by a given species differ also with the latitude—probably because of the different angles of the sun at noon, and hence the varying depths of penetration of the light. Weather conditions, too, have an influence. Heavy cloud will naturally decrease the amount of light that reaches the sea, and zooplankton are found a little higher in the water on cloudy days than on bright days. It seems that even moonlight affects the position occupied by certain species, for they remain farther from the surface on brightly moonlit nights.

Some biologists think that the migrations are controlled almost entirely by light intensity, with limited variations according to temperature, salinity, and other physical factors. They believe that during the day the different species are forced to occupy the depth at which the light intensity suits them best, whereas during the hours of darkness they are able to rise to the feeding grounds near the surface. This theory, however, does not take into account the fact that vertical migrations are also observed among the creatures of the twilight zone, where there is very little light. Nor can it account for the behavior of certain zooplankton species, which, though normally found some distance down, can occasionally be seen swarming at the surface in the brightest daylight.

Although vertical migration is closely connected with the level of light intensity, this is clearly not the whole story. Biologists have found that, under experimental conditions, some animals may make vertical movements that are not directly related to the intensity of light. A copepod called *Acartia*, when removed from the sea and kept in total darkness, continued for several days to change its level with a 24-hour rhythm. This little creature's movements appear to be regulated by an internal "biological clock" that is normally connected with changes in light intensity.

Other theories have been put forward to explain the mechanism of vertical migration. One suggestion is that the rising and sinking of the zooplankton is a simple result of the change in density of an organism according to whether it has recently eaten or not: the animal feeds, becomes heavier and denser, and gradually sinks. Then the food is digested, and the animal becomes lighter and less dense, and slowly rises toward the surface again. This theory is probably an over-simplification, though changes in the density of the zooplankton may play some part in the process.

Why did vertical migration evolve? What advantages do the animals that undergo vertical migration gain? It has been suggested that by remaining in the deeper waters during the day, the plankton animals are better able to escape sharp-sighted predatory fish, and that they rise to the more shallow waters to feed only under cover of darkness. If this is the case, though, why should krill often advertise themselves to any predator as a brilliantly phosphorescent patch on the surface at night? This theory also cannot account for the seasonal and other variations that occur in the migratory patterns.

Another theory is that vertical migration of the animals of the zooplankton may have evolved because of their need to avoid dangerous chemical conditions in the surface waters during the day. Some biologists have suggested that during photosynthesis certain plants of the phytoplankton might give off a substance harmful to zooplankton. There is, however, little concrete

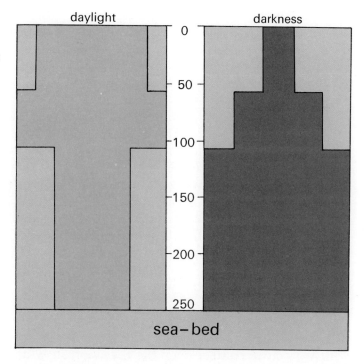

The average relative depth distribution in meters of the diatom Coscinodiscus bouvet near the island of South Georgia in the South Atlantic. The figures show the tiny plants nearer to the surface during daylight hours, exactly opposite to the pattern shown by the zooplankton.

daylight daylight darkness

sea–bed

evidence to support this idea. Indeed, among the creatures that migrate are great numbers of mid-water animals that never feed in the areas where photosynthesis occurs.

Clearly there is no straightforward explanation of the puzzle of diurnal vertical migration. However, when we look at the conditions of life of plankton animals in the twilight zone, we can find some clue as to why they spend the greater part of their time below the surface waters, ascending only for a few hours' feeding at night.

The top 330 feet of water, as well as being the source of all food, has the highest and least stable temperature almost everywhere in the world. Below this layer is a zone known as the *thermocline*, in which the temperature drops sharply. Beneath this zone it falls only very slowly. (The only exceptions to this rule are in the North and South polar waters which are coldest at the surface because melted ice reduces salinity, and in high-salinity basins, such as the Red Sea.) During the day, at least, the animals of the twilight zone are living in water that is much cooler and more constant in temperature than that of the surface layer. This evenness of temperature over most of the globe provides a steady environment, which means that there is a vast amount of living space for animals adapted to a fairly low temperature of about 46–50°F.

Why is a low, stable temperature an advantage? Briefly, in an animal such as a fish or a crustacean, whose body temperature varies with the temperature of its environment, its rate of *metabolism* (energy-expenditure) is speeded up by a rise in temperature and slowed down by a fall in temperature. The rate of metabolism increases 2–3 times for every 10°C rise in temperature, and decreases by the same amount with a similar fall in temperature. This explains the advantage for marine animals of living below the thermocline. If, as in the waters of the North Atlantic Ocean, the surface temperature is an average of 55°F and the twilight zone animal normally lives at 45°F, it only needs to eat about half as much food as it would if it remained at the surface.

Thus the creatures of the twilight zone have the double advantage of living economically and at the same time close to their main food supply. Furthermore, those animals that do not actually enter the productive zone are well placed to browse on the constant rain of dead and dying plankton from above. For small, slow-moving animals the twilight zone is also a safer habitat than the photic zone, which is the home of many active predators, especially fast-moving fish. There is one last advantage: the colder the water, the greater its density and viscosity. This means that twilight zone plankton animals need to spend less energy in maintaining their positions than they would if they lived in the warmer surface waters.

Animals that make vertical migrations gain another extremely important advantage. Generally speaking, plankton is helplessly adrift in the sea, with very little control over the direction in which it is carried by the currents. A vertical migration, however, can carry an animal from water moving in one direction to water moving in another direction—or at least moving at a different speed. In fact, water masses are hardly ever moving at

Right: Vertical migration in layers of water swirling in opposite directions can keep animals in one locality, as here, where two animals move from A to A1, and B to B1 and back. Below: Two different flows can result in a horizontal change of position for plankton.

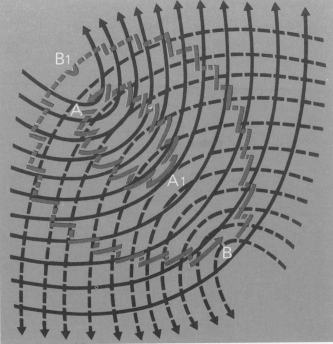

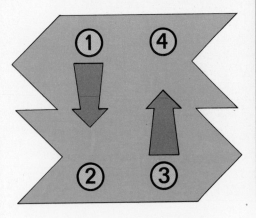

the same speed at different depths. Wind action naturally has the effect of moving the surface water faster than that of the depths below, and in comparatively shallow waters, the deeper water is also slowed down by friction with the bottom.

Whether the surface layer is moving faster but in the same direction as the water below, or whether it is moving in a different direction, 24 hours can easily bring about a change of half a mile or more in the relative position of two water masses, one at the surface and one below. Thus, if a plankton animal can move upward or downward through 300 feet or so, into a body of water moving in a different direction, its position can be changed in one day by half a mile or more. This change of environment is important because it gives the plankton animal more chance of finding food by moving on to fresh pastures. Also it enables the larvae of plankton animals to be dispersed widely. How important these movements are to the plankton can be seen by examining the region of the Antarctic Convergence. A look at the pattern of these Antarctic currents will help toward our understanding of the vertical migration of plankton.

The South Atlantic has been explored over a period of 26 years in more detail, perhaps, than any other deep-sea area, by the ships of the British *Discovery* expeditions. From these deep waters, free from tidal streams, oceanographers have built up a clear picture of the movements of the ocean currents all the way from the surface to the bottom. At the Convergence the cold Atlantic surface water flowing north meets the warmer surface water of

Above: The crustacean Cystosoma, *which lives in the twilight zone of the ocean. Its minute body is crystal clear, allowing an observer to watch its internal organs.* Cystosoma *has been caught in nets as deep as 12,000 feet below the surface.*

the sub-Antarctic region. Where these currents meet, the Antarctic surface water dips beneath the sub-Antarctic surface water because the latter, though more saline, is warmer and on balance less dense. From the Convergence, the water that has dipped beneath the sub-Antarctic surface current—the Antarctic Intermediate Current—moves slowly northward as far as about latitude 30°N (the latitude of Florida and the Canary Islands) where it gradually merges with the surrounding waters and becomes indistinguishable as a separate current.

When we look at the pattern of currents just south of the Antarctic Convergence, we see that the Antarctic Surface Current is comparatively shallow, being no more than about 330 feet deep. Beneath it, the North Atlantic Deep Current flows *south*, all the way to the ice-edge.

To find out how this affects the migration of plankton we must reconstruct the conditions of life for a plankton animal that feeds in the highly productive waters of the Antarctic Surface Current. It is summertime and there is food in abundance in the photic zone—plenty of phytoplankton for a grazer, and plenty of zooplankton for a predator. But all the time, the little plankton animal is drifting north while it feeds, and gradually approaching the Antarctic Convergence where it seems certain that it will be carried down into the Antarctic Intermediate Current, below the photic zone, and ultimately to its death. But if it makes a daily migration downward into the south-going North Atlantic Deep Current, returning to the Antarctic Surface Current during the night-time, it can to some extent stay in the same geographical position.

Here, then, is another example of the survival value of diurnal vertical migration. Nevertheless, although the plankton animals can stay longer in the Antarctic surface water by means of these daily migrations, their rate of travel in the Warm Deep Current does not *quite* make up for the effect of the northern drift at the surface. One might therefore think that the plankton *must* eventually be carried on into the Antarctic Convergence. The long-term result of this would inevitably be that the Antarctic waters would be stripped bare of their plankton population, and hence of every other form of animal life. But we know that this does not happen.

What, then, prevents the depopulation of the Antarctic waters? The answer, it seems, is a *seasonal migration*. Such a migration has been observed in this area among plankton animals that drop down into the North Atlantic Deep Current at the beginning of the southern winter (in March) and remain there, to be carried southward, throughout the winter. The diagram opposite shows the results of a series of samples taken at different seasons of the year by the research ship *Discovery II*. It can be seen from the sample taken at the onset of winter that a vast body of mixed plankton has sunk into the North Atlantic Deep Current and is traveling south. As the North Atlantic Deep Current drifts south, it is all the while collecting debris —dead and decaying plankton from the surface and, more important, from the northward-drifting Intermediate Current. We say "more important" because a large proportion of the plankton does not return to the ice edge and must presumably perish either at the Antarctic Convergence or at the next area of marked change in conditions—the Subtropical Convergence, some 620 miles farther north.

The North Atlantic Deep Current finally rises to the surface in the far south, mixes with melt-water from the polar ice, and starts a new journey northward as the Antarctic Surface Current. And so we have come full circle, back to the beginning of the cycle. After having traced it right through, we can see how the new surface water is loaded with nutrients and with eggs, spores, and larval forms of plankton. Only one thing is needed to trigger off a new explosion of plant and animal life, and that is sunlight. And in the high latitudes of the Antarctic, the extremely long summer days fully make up for the lack of intense sunlight.

It is likely that cyclical migrations over long distances are made by all Antarctic plankton animals, and by phytoplankton in certain resting or spore stages. If this were not so, the extreme southerly waters would not remain stocked with such an abundance of plankton.

Although many of the problems concerning vertical migration are still unsolved, it is clear by now that the phenomenon is of very great importance to marine life, not only on a local scale, but also over long periods and extremely wide areas.

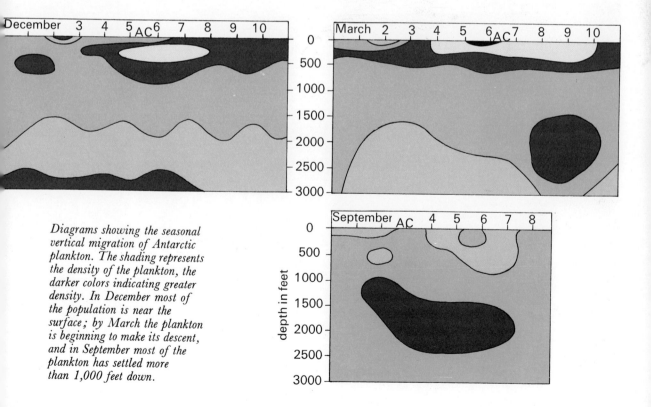

Diagrams showing the seasonal vertical migration of Antarctic plankton. The shading represents the density of the plankton, the darker colors indicating greater density. In December most of the population is near the surface; by March the plankton is beginning to make its descent, and in September most of the plankton has settled more than 1,000 feet down.

5

The Sunless Deeps

Because we are so used to the rhythms of day and night, winter and summer, it is very difficult for us to imagine what life must be like for the creatures of the ocean depths. Below about 2,000 feet there is no daylight, but only perpetual darkness; no summer warmth, but only perpetual cold; and the deep ocean currents flow so slowly that the movement is scarcely noticeable.

It seems incredible when voyaging across the Equator, where the sun beats down on sparkling blue water, that a mere 2,000 feet below the keel there is utter darkness, silence, and a temperature of only about 7°F above freezing point.

In most areas of the ocean the temperature falls dramatically with increasing depth. The contrast between the warmer surface waters and the cold deep waters is naturally sharpest in tropical and subtropical regions. Yet once the depths are reached, temperature varies little from one geographical area to another. At 6,000 feet the water is at an average of 37.4°F in the North Atlantic and 39.2°F at the Equator. The lowest temperature at the bottom of the ocean along the Equator is about 36.5°F. However, the temperature at the sea floor does become somewhat lower toward the south, until in the Antarctic the bottom temperature is about 31°F. (This water does not freeze because of its high salinity.) The ice at the South Pole influences the oceans even at great depths, so that the deep South Atlantic is colder than the deep North Atlantic.

Everywhere, then, the creatures of the ocean depths live in a still, almost unchanging world—a world where there is perpetual night, there are no seasons, and little water movement. It might be expected that because of this evenness of habitat most deep-sea animals would have a worldwide distribution. In fact, although there is much that we do not know about deep-sea life, it appears that the animals there are quite localized, and there are few actually worldwide species.

We know that the top 650 feet are the only primary source of food, so how can life exist at great depths? A parallel can be drawn here between the oceans and a tropical rain-forest on land, where most of the

A deep-sea squid, Calliteuthis, *which lives in the dark, cold ocean between 1,500 and 4,500 feet down. The underside of its body and its tentacles are studded with brilliant light organs, called photophores, a bright light in the blackness.*

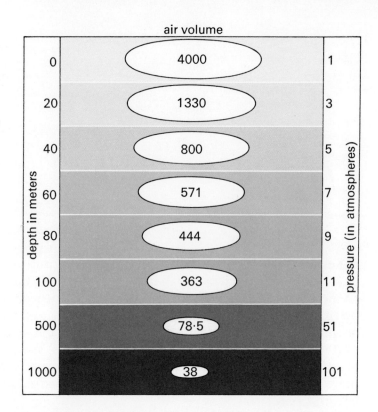

The changes in the volume of a fish's swim-bladder at varying depths. The pressure shown at the right is the combined hydrostatic and atmospheric pressures at that depth. The volume changes are less the deeper the fish lives, and so the deep-sea fish can rise and descend over a greater distance without having to make abrupt corrections to the volume of gas present in its swim-bladder.

biological activity is in the treetops. The foliage of tall trees, where there is great competition for sunlight, is the area of primary production—the counterpart on land of the photic zone in the sea. The debris from among the treetops—dead animals and leaves—piles up on the ground and is eaten by animals or decomposed by bacteria and fungi. The nutrients (mainly phosphates and nitrates) are finally recycled into the roots of the trees, and up into the productive foliage again.

In the deep oceans there is a similar pattern. Most of the organic matter is produced and used up at the surface, but some of this food is used directly and indirectly right down to the bottom. There is a steady reduction in the amount of plankton all the way down, as it is eaten by predators. At 3,000 feet the plankton is only one-tenth as plentiful as it is near the surface. Below 3,000 feet the number of organisms falls off more slowly but steadily, until just above the ocean floor there are very few indeed. Most of the food supply in these deep waters consists of detritus from above. As this detritus continues to sink it is being broken down by bacteria, so that its mass becomes smaller all the time, and only a fraction reaches the seabed. This is why there can be no teeming millions of creatures in the deep waters.

In the ocean depths biologists separate two main zones: the intermediate or *bathypelagic* zone and the deep or *abyssal* zone. Just as the creatures of the upper waters are specifically adapted to life either in the photic zone or the twilight zone, so the creatures of the depths are specifically adapted to life either in the bathypelagic or the abyssal zone. Nevertheless, *all* deep-sea creatures share certain conditions in common, though those conditions vary in degree. The most important ones are great pressure from the weight of water above, and darkness.

The pressure that water exerts, on itself and on any object submerged in it, is called *hydrostatic pressure*. But we have to take both atmospheric and hydrostatic pressure into account when looking at pressure beneath the waves.

The atmosphere presses on our world with an average weight of 14.7 pounds per square inch. Each of us is living at the bottom of a column of air that presses on our bodies with a total load of about 10 tons. Because we are adapted to life at this pressure, we are quite unaware of the load that presses on us equally from all directions, and from the pressure inside as well as outside. But if it pressed on one side only, we should be squashed flat. The main internal air space of the body, the lungs, is always in contact with the atmosphere (except for a fraction of a second when we swallow), so that the pressure inside and outside the ribs is the same. Another air space inside the body is in the middle ear, and here it is important that the atmospheric pressure on each side of the delicate eardrum is equal. The connection between the middle ear and the atmosphere is by way of the *eustachian tube,* which leads to the back of the throat. The eustachian tube normally remains closed, but it opens for a moment every time we swallow naturally, and this is quite enough to take care of ordinary variations in atmospheric pressure. This is why the unpleasant sensation in the ears felt when taking off or landing in an aircraft (caused by sharp changes in pressure) can be relieved by swallowing frequently.

It is absolutely vital for a diver to equalize internal and external pressure, because water is about 800 times denser than air at sea-level, so that as he descends the *additional* external pressure will increase rapidly. The word additional is emphasized, because at the surface the diver's lungs fill with air at atmospheric pressure, and when he dives hydrostatic pressure is added to this.

Boyle's Law (named after Robert Boyle, 1627-1691, the British physicist who first discovered it) states that when a gas stays at the same temperature, the volume of the gas decreases proportionately as the pressure increases. And when the pressure decreases, the volume of gas *increases* proportionately.

If a diver takes a fairly deep breath, so that his lungs contain a total of 4,000 milliliters of air at atmospheric pressure, and then submerges quickly, holding his breath, by the time he reaches a depth of 30 feet, the external pressure acting on his lungs is 2 atmospheres (one for the atmospheric pressure and one for the hydrostatic pressure at that depth). Now the diver's lungful of air has been compressed to 2,000 milliliters—half what it was at the surface. In other words the gas has behaved according to Boyle's Law: when the pressure is doubled, the volume is halved.

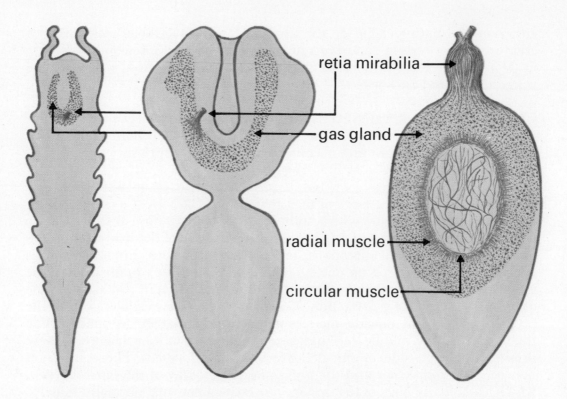

retia mirabilia

gas gland

radial muscle

circular muscle

If the diver now drops another 30 feet, to a depth of 60 feet, he will be under a total pressure of 3 atmospheres—three times what it was at the surface. The volume of air in his lungs will then be only 1,333 milliliters—one-third of the original volume. Again, true to Boyle's Law, when the pressure is trebled, the volume is reduced to one-third of its original value.

But what has Boyle's Law to do with fish? Most fish chase their prey, using up a great deal of energy in the process. Fish are active and agile swimmers, and have a sturdy skeleton and a robust muscular system, both of which make their solid structure about 5% denser than water. If a fish had no means of controlling its buoyancy, it would continually have to waste energy merely to hover in the water, and to do this it would have to exert a continual upward force to about 5% of its own weight. Many fish escape this problem, however, because they have a *swim-bladder*—a sort of gas-filled buoyancy tank.

In general we can say that if a marine fish is to float without effort it has to have a swim-bladder with a capacity equal to about 5 percent of its volume. The swim-bladders of some freshwater fish have a larger capacity—about 7 percent of the volume—because fresh water is less dense than seawater and therefore supports proportionally less weight per volume.

As we can see from the diagram above, a fish whose range is within the top 30 feet of water undergoes a bigger change of swim-bladder volume—50

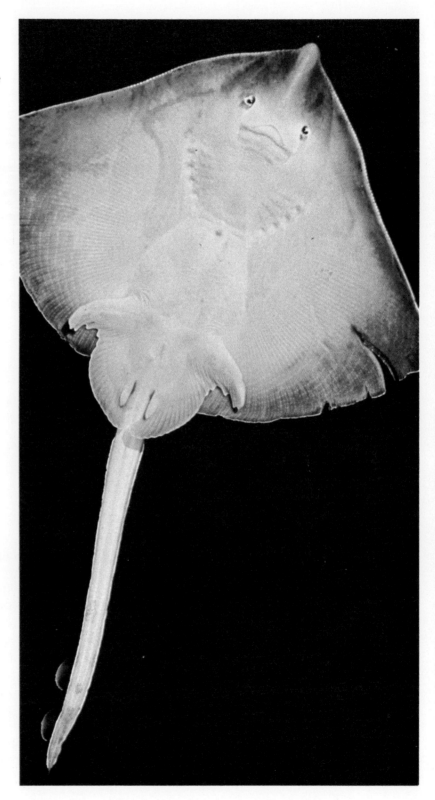

Left: Diagrams of the swim-bladders of fish. The two on the left, both 4½ inches long, are of shallow-living fish, members of the cod family. The one on the right, only ¼ inch long, is of Yarrella blackfordi, *a deep-living fish. The gas-producing tissues are much more extensive in the fishes of deep waters.*

Right: A thornback ray, Raia clavata, *a member of the large class of fish that includes sharks. These fish have skeletons composed entirely of cartilage, which is lighter than bone. In addition, their livers contain low-density substances that also help in making them buoyant.*

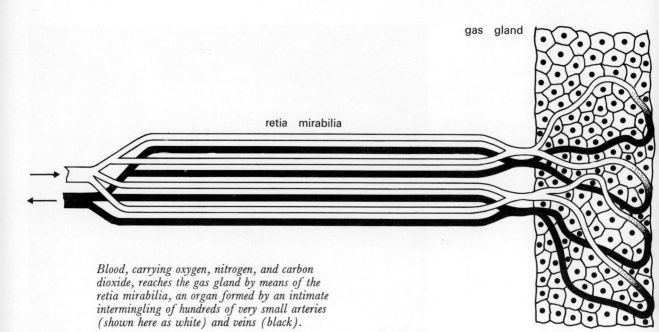

gas gland

retia mirabilia

Blood, carrying oxygen, nitrogen, and carbon dioxide, reaches the gas gland by means of the retia mirabilia, an organ formed by an intimate intermingling of hundreds of very small arteries (shown here as white) and veins (black).

percent—than a fish moving through the same vertical distance in deeper water. This is because of the difference in pressure (and hence of the volume of the swim-bladder) with depth. Fish that live in the surface waters need a greater degree of adjustment in their swim-bladders than those living in deeper waters.

Most freshwater fish are surface-water dwellers, and have an open swim-bladder—that is, the bladder has an air duct connecting it to the front end of the gut. The fish becomes buoyant at the surface by swallowing air with the aid of a swelling on the air duct called the *pneumatic bulb*, which pumps up the swim-bladder. As soon as the fish submerges, however, the volume of the swim-bladder diminishes under pressure. The fish would therefore no longer be buoyant if it were not for a *gas gland* that extracts dissolved oxygen and nitrogen from the fish's blood and secretes it into the swim-bladder, thus "topping it up" to the correct volume to achieve buoyancy. Suppose now that a freshwater fish wants to surface quickly from a depth of 30 feet. If there were no duct, the gases would expand to double their volume and the fish would literally burst. But, the open duct allows the excess gas (secreted by the gas gland) to escape from the mouth in a stream of bubbles.

By contrast, most marine fish have closed swim-bladders (except for the herring family and the eels). Dolphins and tuna, high-speed predators of the surface waters of the ocean, either have a small swim-bladder that is closed, or none at all. This may be because their oily flesh allows them to be very nearly buoyant even without a swim-bladder. It may also be that a swim-bladder would be a positive hindrance to them, since these fish continually submerge and resurface in pursuit of their prey in just that region of the sea where the greatest proportional change in gas volumes takes place.

There is yet another reason why swim-bladders may not have been evolved in tuna. Just as the gas-filled spaces within plankton creatures can bounce back an echo to an ultrasonic depth finder apparatus on board a ship, it seems likely that swim-bladders would act in the same way. Now tuna are preyed upon by killer whales—marine mammals that have their own natural ultrasonic depth-finding equipment that they use to locate their prey. It is possible, therefore, that tuna have not evolved swim-bladders because these would be a disadvantage to the fish, enabling the whales to catch them more easily.

Other fish with no need for a swim-bladder are the many bottom dwellers of shallow waters and of the continental shelf, down to about 650 feet. Most familiar of these are flatfish, such as sole, plaice, halibut, and flounder. All these are denser than seawater, so that even if they do leave the bottom for a while they automatically sink gently back to the seabed, which is their feeding ground and their home. Some members of the shark family—for example, the rays—though more active predators, also tend to sink back to the bottom when they are inactive.

In those fish that live in the twilight zone, down to about 3,200 feet (though not those found on the seabed at this depth), the swim-bladder again plays an important part. Many of the common *pelagic* (open sea) fish, such as lantern-fish and hatchet-fish, have well-developed closed swim-bladders. These fish have been traced by sonic depth finders as they move upward by night and down again by day, through as much as several hundred feet of water. So it is clear that their means of buoyancy adjustment must be very effective. It is in fact provided by a well-formed gas gland consisting of many layers of cells covering a large part of the inner surface of the swim-bladder. This gas gland is supplied with blood by the *retia mirabilia* (Latin for wonderful net), a complex of hundreds of blood vessels. This is shown in the diagram opposite.

In the sunless waters below the twilight zone are fish that have no swim-bladder. They are lightly built, having little bone or muscle, and this delicate structure makes them buoyant. These fish do not hunt actively, but hover in the water, enticing their prey into their large mouths by means of various appendages that serve as lures. Angler-fish and gulper-eels are typical examples. There are also inhabitants of the abyssal seabed, tripod-fish for example, that have no swim-bladder.

On the deep-sea or abyssal floor, however, at depths from about 6,500 to 23,000 feet, are found more active and more sturdily constructed fishes, such as rat-tails and halosaurs, that *do* have swim-bladders that can work under great pressures.

Yet the swim-bladder is not the only buoyancy device in fishes. All fish of the shark family have skeletons that are composed entirely of cartilage, a material lighter than bone. In these fish buoyancy is further increased by the

presence of low-density substances, stored in the liver. Even so, some of these fish still remain slightly heavier than water.

Fish are not the only animals that can control their depth in the water. Some invertebrates, too, have evolved methods by which they can control their buoyancy, such as the Portuguese man-of-war (*Physalia*) and the by-the-wind-sailor (*Velella*) that have parts of the body modified as sails that stick up above the surface of the sea (Chapter 3). Both *Physalia* and *Velella* contain special gas-secreting cells and belong to the group of animals called the siphonophores.

But there are also underwater siphonophores with limited powers of swimming, and these, too, can adjust their buoyancy at different depths. They do this in various ways. One is by means of a large *somatocyst*, an oily structure that counterbalances the tendency to sink. The siphonophore called *Diphyes appendiculata* has not only a somatocyst, but also jelly-like swimming bells. The salt content of these swimming bells can, it seems, be altered, thus bringing about a change in density.

Octopuses, cuttlefish, squids and their relatives have particularly interesting systems for altering their bouyancy. One of the most intriguing is that of the common cuttlefish that lives in coastal waters and ranges from the surface to about 300 feet down.

By day this creature rests on the bottom, hiding in the mud and sand. By night it hunts near the surface, swimming around by squirting jets of water behind it. Completely inside the cuttlefish's body is a shell or *cuttlebone*, made of horny matter covered with a chalky layer. Without this cuttlebone, the cuttlefish's body would be denser than seawater, making its nightly rise to the surface very difficult. But the density of the bone is much less than that of water, and so gives the animal buoyancy. Equally important, the bone's density can be increased, lessening the cuttlefish's buoyancy and allowing it to sink easily by day.

The bone is full of many small chambers containing both liquid and gas. When the cuttlefish needs to increase its buoyancy, it removes salt from the liquid in the chambers. This causes the liquid to pass into the more saline bloodstream by the process of *osmosis*. (This is the process by which water from a weak solution of salt passes into a stronger salt solution across a membrane between them). As liquid passes out of the chambers of the cuttlebone, the gases in the chambers expand, filling the empty spaces and making the whole cuttlebone less dense. The cuttlefish is then most buoyant, and can stay up at the surface. When the cuttlefish returns to the bottom at daybreak, fluid re-enters the chambers, decreasing its buoyancy and helping it to stay on the bottom.

A cuttlefish that lives in rather deeper water in tropical and subtropical regions is the little creature called *Spirula*. Young ones live at about 3,300-5,700 feet, but adults prefer shallower water, at a depth of about 650 feet.

Above left: A photograph of a cross-section of the shell of a nautilus, called the pearly nautilus. Below left: A cuttlefish Spirula. With nautilus, it is a survivor of a group of invertebrate animals that flourished 200 million years ago.

Below right: A drawing of nautilus, showing the porous siphuncular tube. Like a wick, it draws up fluid in the chambers, altering the buoyancy. The animal itself lives in the outermost chamber, and when it swims, that chamber is downward.

This animal has an internal spiral shell divided into a number of gas-filled chambers, which are each completely enclosed. These buoyancy chambers lighten the animal and more than make up for the weight of the shell. They also enable *Spirula* to regulate its buoyancy in much the same way as does the larger common cuttlefish.

Squid, which live in deeper water than cuttlefish, have within their bodies a very small bone called a pen, which offers little buoyancy control. These creatures have to swim actively to stay afloat, like the mid-water fish of the bathypelagic zone that have no swim-bladders.

Life in the sea ranges from shallow coastal waters to the deepest ocean trenches. The floor of the Challenger Deep, 35,750 feet down, where the pressure is 14,000 pounds per square inch, is inhabited. There, in the bathyscaphe *Trieste*, Jacques Piccard and Don Walsh saw a flatfish and a red shrimp, and the ocean bed showed signs of the tracks left by other creatures. The Danish research ship *Galathea* photographed a brittle star and grabbed a sample of ooze containing bacteria from the bottom of the Philippine Trench, 32,997 feet deep. Samples of the bacteria were later grown on board in a special apparatus for applying the same pressure as in the natural habitat. After three days at 100 atmospheres and a temperature of 37°F (the approximate temperature at the bottom of the Philippine Trench) they grew profusely. Yet attempts to culture the same bacteria at atmospheric pressure

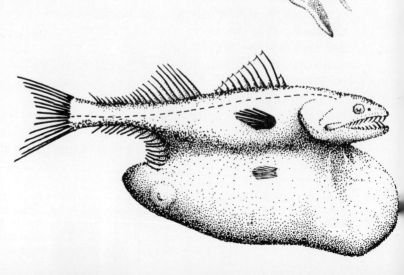

Above: The common cuttlefish, Sepia officinalis, *which can change its buoyancy by varying the density of its shell, the cuttlebone. It lives in coastal waters, moving upward at night.*

Right: Chiasmodus, *the "Great swallower," a fish only four to six inches long that can swallow other fish considerably larger than itself. This little fish lives in the deep ocean regions, between 1,500 and 4,500 feet down. There, where prey is scarce, it is an obvious advantage to be able to consume any available food, even if it is bigger than the voracious predator itself.*

failed. Organisms that flourish under high pressure are called *barophilic* (Greek *baros*—weight, *philos*—loving).

Vertical migration almost certainly ceases below the twilight zone, and the deep-sea plankton and nekton do not reach the upper layers. How, then, do deep-sea fish and other marine animals find enough food?

In fact the difficulty is not as great as one might suppose. In contrast to the comparative barrenness of the middle or bathypelagic waters, the ocean bottom is rich in living matter. It is the final resting place of all the debris that has escaped being consumed, decayed or dissolved on its long descent to the abyssal plains. As a general rule, the deeper the sea the less debris survives to enrich the bottom mud.

The sediments of the deep ocean-bed are made up of detritus—a mixture of organic and inorganic particles. There is also a large population of bacteria, and in the top few inches of the oozy mud these may be as numerous as they are in fertile garden soil, with several million bacteria per gram. It is not surprising, therefore, to find that the principal animal inhabitants of the deep-sea floor are mud eaters—sea cucumbers, brittle stars, starfish and sea urchins. These live in much the same way as earthworms—by stuffing their guts with ooze to take in the bacteria and perhaps also dissolved organic compounds.

There are other animals living on the abyssal seabed that get their food by filtering it out of the water in various ways. Filter-feeding sea anemones have been found on the floor of the Philippine Trench, together with other related animals, such as the plant-like sea pens and gorgonians. Many species of filter-feeding worms also live on the deep-sea floor, and most of them hide within tubes that they build from the shells of dead animals and from grains of silica. Some mollusks, too, filter particles of food falling from above, and there are huge glass-like sponges, some as high as three feet.

Crawling across the deep-sea floor are other animals, mainly crabs, prawns, and sea spiders. Some seem to shovel up the ooze and separate out particles of food, while others are predators or scavengers.

All these bottom-living animals are preyed upon in turn by deep-sea fish. Rat-tailed fish, for instance, swim near the bottom and probably root about in the oozy mud searching for their burrowing prey. Remains of worms, crustaceans and pieces of sponges, all of types that live on the abyssal seabed, have been found in the stomachs of these fish. Abyssal fish do not feed only on bottom-living animals. They also swim above the seabed, catching other fish and invertebrates in the water.

So the deep-sea floor is by no means a lifeless desert. Compared with the mid-waters, it is quite fertile.

Several species of animals whose presence in large numbers can set the surface waters glowing brightly by night were described in Chapter 3. In the darkness of the deeps, too, live many animals that produce their own

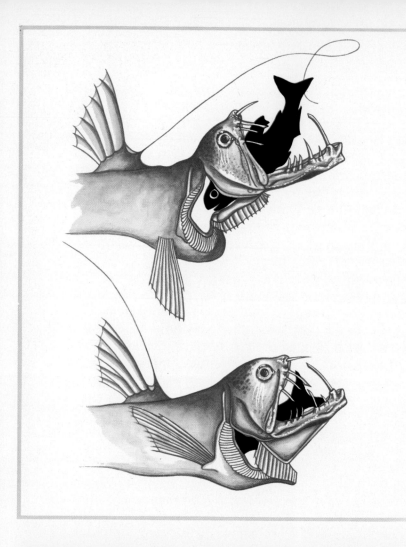

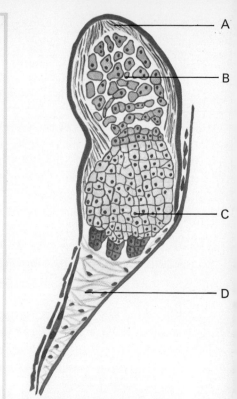

Above: A diagram of a cross-section through a photophore of the hatchet fish Argyropelecus affinis. *A is the reflecting layer, B the mass of cells that produce the light, C the color filter, and D is the lens.*

Above left: Chauliodus, *the viper fish, swallowing a lantern fish. The viper fish lures its prey by an angling device that is a modified ray of the dorsal fin, tipped by a light organ that can be dangled temptingly in front of a fish. To swallow such large prey, Chauliodus throws back its head, which pulls down the lower jaw so that the bones supporting the gills are pulled out of the way.*

Left: A photograph of the head of a viper-fish, showing the amazingly movable jaw.

Right: The photophore of Ctenopteryx, *a deepsea squid. These organs are used mainly for recognition or attracting prey in the dark.*

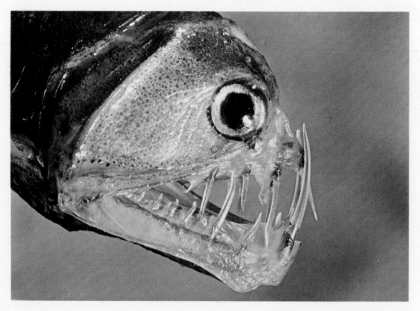

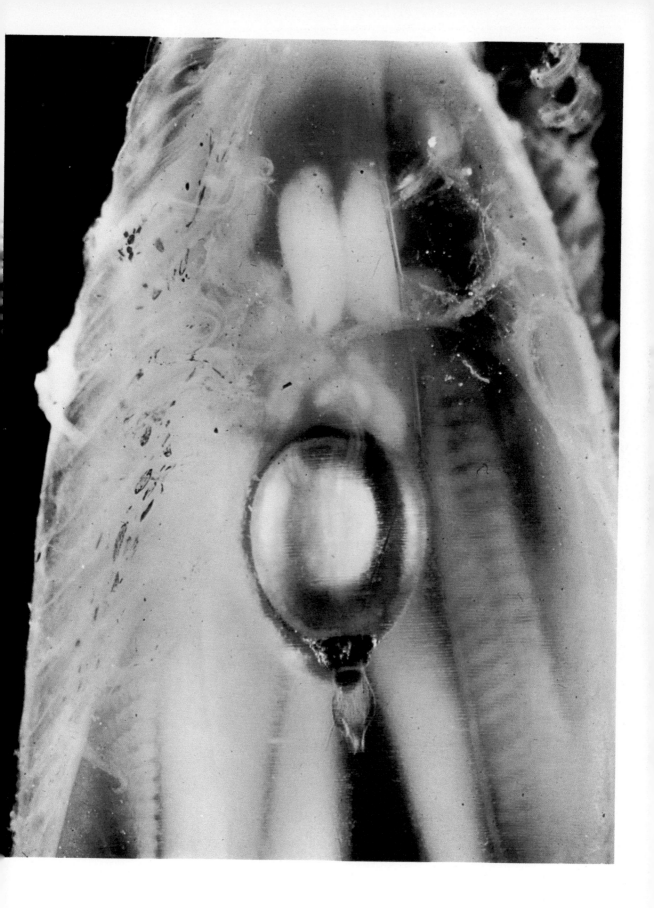

light. Living lamps of the dark waters are to be found among squids, deep-sea prawns, and a majority of twilight zone fish. And some animals of the seabed that live fixed in one place, such as gorgonians and sea pens, have been seen with every detail of their elegant structures glowing with luminescence.

The production of light by living things is called *bioluminescence,* and it is brought about by the interaction of two chemical substances called *luciferin* and *luciferase.* Bioluminescent light is sometimes called "cold light," because the energy it radiates is almost entirely within the visible spectrum. Only one percent of the energy that goes into its production is lost in the form of heat. Compare this with what happens when we convert electrical energy into light. Even the best we can do incurs a heat loss of 50 percent!

Most bioluminescent animals produce light by means of special organs called *photophores,* made up of an assemblage of light-producing cells, but light can also be produced by luminous bacteria living in association with the animal. Photophores seem to have evolved independently in different animals but their structure may be quite similar in such widely differing groups as fish, crustaceans and squids.

Fishermen have known for hundreds of years that marine animals are attracted by light—not by the general illumination of daylight but by individual sources of light. The Mediterranean fishing fleets can be seen twinkling at night over miles of sea as they burn gas flares to attract shoals of fish into their nets. Fishermen in various other parts of the world use luminescent material from bioluminescent fish to help increase their catches. For instance, natives of the Banda Islands in Indonesia bait their hooks with a gland taken from a fish called *Photoblepharon.* This gland contains luminous bacteria that shine for several hours. Fishermen at Sesimbra, in southwest Portugal, use as a bait a piece of strong-smelling dogfish smeared with the luminescent yellow fluid that exudes from a gland in the belly of a rat-tailed fish. The fish they catch are attracted by both the smell and the light.

There seems little doubt that many bioluminescent fish use their light just as human fishermen use it—as a form of bait or lure. Some deep-sea fish have luminous organs on the inside of their mouths, and so are able to attract small crustaceans and fishes into the brightly lit cavern.

One of the angler-fish, that lives from about 3,000 feet downward in permanent darkness, goes fishing complete with rod, line, and luminous bait. The female has a "fishing-rod" mounted on a bone controlled by muscles, and this can be pulled backward or forward in a groove along the fish's back. At the end of the rod is a luminous organ that can be switched on and off. It seems that when the angler-fish is searching for its prey, the rod is in the forward position, so that the lure dangles in front of its mouth. When an inquisitive victim comes to investigate the light, the rod probably slides back, bringing the light closer to the mouth. Then the line is whipped away and the victim is swallowed.

Only the female angler-fish behaves in this way. Both male and female

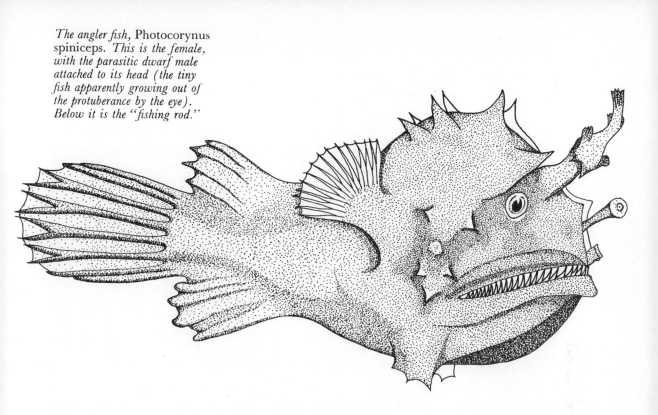

The angler fish, Photocorynus spiniceps. *This is the female, with the parasitic dwarf male attached to its head (the tiny fish apparently growing out of the protuberance by the eye). Below it is the "fishing rod."*

hatch from eggs, which, as far as we know, are laid in deep water and then float to the surface. As young ones, male and female are very similar in appearance until they are about $\frac{1}{2}$ inch long. Then, as time passes, the female grows up to three feet long, and begins to develop the fishing-rod. But the male remains small, hardly ever exceeding six inches in length, even when adult. It develops large eyes and special protruding teeth called *denticles,* but no fishing rod.

After both male and female have migrated to the darkness of the depths, the male uses his well-developed eyes to find a mate. Here bioluminescence plays the role of cupid, for the little male probably "homes" on the female's fishing light. He then attaches himself to her permanently, by means of his sharp denticles. His blood vessels and those of the female become closely associated, and for the rest of his life he depends on her for nourishment. In return he will provide the necessary sperm to fertilize her eggs during the spawning season.

Among the great variety of illuminating organs found in marine animals, some are definitely in the same class as an automobile's headlights. They produce a powerful light that illuminates the water ahead of or below the animal. In some cases the natural headlight is made up of three parts: a photophore to produce the light, a reflector behind it to throw the light forward, and—in front of the photophore—a lens to focus the light beam. Euphausiids and a number of fish that live in the twilight zone have a light organ incorporated into the eyeball itself. As the animal swivels its eyes in search of food, the light follows the path of vision.

Even more extraordinary is the small subsidiary light organ of some midwater fish such as the hatchet fish. This "secondary lamp" seems to serve only one purpose—to shine light directly into the eye of the fish that carries it. While nobody can be quite sure what its use is, the most likely explanation is that the extremely sensitive eyes of these fish would be dazzled by their own main "headlights." Indeed, the secondary lamp, or *orbital illumination*, may prepare the eye for a later flood of light from the main photophores.

When a cuttlefish is attacked, it squirts out a dense cloud of ink. In daylight in its shallow-water home, this cloud acts like a smokescreen, enabling the cuttlefish to make its escape. But how can an animal that lives in the darkness of the deeps avoid attracting attention to itself when it shines a light to seek or entice its prey? In this case, the animal can use a cloud of *light* to confuse its enemy. There is a squid that can discharge a sticky secretion from two glands that open near its mouth. In the water this secretion turns into a cloud of sparkling lights that confuse the attacker while the squid turns off its own lights and makes its escape. Some deep-sea angler fish and prawns can produce similar confusing clouds of light in an emergency.

One other use of light organs by marine animals, shown in the diagram below, is to help them recognize members of their own species.

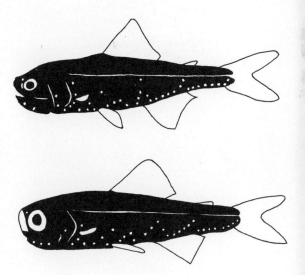

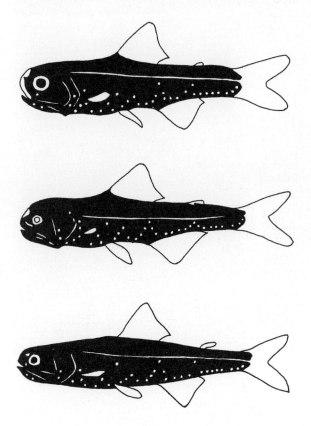

Above: The different patterns of light organs in a closely related group of lantern fish. In the left row, from top to bottom. Diaphus macrophus, Diaphus lucidus, *and* Diaphus splendidus, *Right row,* Diaphus garmani *and* Diaphus effulgens. *Each species has developed its own particular pattern, different from all the rest, which enables a lantern fish to identify other individuals of its own species— an obvious necessity in the dim light of twilight zones in the oceans.*

Left: Two hatchet fish, Argyropelecus hemi-gymnus. *These little fish, about two inches long, live in the twilight zone, between 750 and 1,500 feet deep. Their large tubular eyes give wide-angle binocular vision looking either upward or forward.*

6

The Continental Shelf

The earth's crust varies greatly in thickness from place to place. The ocean basins, formed of dense, dark, basalt rock, are areas where the crust is relatively thin. The continents, lying over the basaltic layer of the crust and formed from the far less dense rock called granite, are areas of greater thickness. Water fills the ocean basins to overflowing, and drowns the edges of the overlying continents to varying degrees. The submerged continental margin is known as the *continental shelf*. Throughout the earth's long history the area of the continental shelf has varied from time to time, depending partly on earth movements, but also on the sea level, and today's coastlines— the familiar outlines on a map of the world—merely represent a temporary state in the constantly changing level of the sea.

There is a good deal of evidence from both the recent and the remote geological past of the rising and falling levels of the sea. For example, during the last of the four great Ice Ages, or glaciations, which was coming to a close about 20,000 years ago, colossal quantities of water evaporated from the ocean and became locked into ice on land. This so greatly reduced the amount of water in the oceans that the average world sea level fell by nearly 500 feet, exposing much of the continental shelves. Today many of the areas exposed at that time are once again under water, and around the coastlines we find the remains of *submerged forests*—a reminder that these areas were once dry land.

While the sea level has risen in some places since the end of the last Ice Age, it has fallen in others. This is because parts of the northern continents, formerly weighed down by ice sheets, are now in the process of readjusting themselves, and are rising slowly. Thus we find *raised beaches* around many coasts in Europe and North America. Some of these lie only a few yards above today's mean sea level, a little way back from the present sea line, but others are to be found high on sea cliffs. They contain the remains of countless shore-living mollusks, and these remains are often of the same species as those being washed up by the waves on the seashore today.

If we want to go back further in time to show how the pattern of land and sea has changed, we have only to look at the world's great mountain chains.

Heavy seas pound the coastline near Estoril, in Portugal. Around the world the waves are carving changes in the rocks and sand they ceaselessly rush up against, both by the impact of the crash and the suction of the retreat back to the sea.

Most of the world's major mountain ranges are composed of rocks that were once laid down as sediments on an earlier continental shelf.

Looking back still farther, about 500 million years ago, the continents were all grouped together in one great landmass. This "supercontinent" split into pieces that drifted very gradually to their present positions.

Yet although continental shelves are slowly but constantly changing their boundaries, they always have certain features in common. They have very gentle gradients (about 1 in 500) from the shoreline to a point about 400 feet below sea level. Then the gradient changes quite abruptly to an average slope of 1 in 15. At this point the *continental slope* begins, and this runs almost unchanged in gradient right down to the deep ocean floor, usually being cut into by underwater valleys in many places.

The shelf underlies 7.5 percent of the total area of the oceans, which is equivalent to 18 percent of the total dry-land area of our planet. And this vast shallow-water region is the most productive part of the oceans. The reasons for its high productivity are many, and in this chapter we shall look at the physical conditions that make it so.

A river carries material down to the sea in three ways. It carries a *dissolved* load of chemicals, a *suspended* load of fine rock particles and scraps of organic material, and a *bed* load of coarse debris and stones that it rolls along the riverbed.

The faster a river flows, the larger is the bed load of large stones, and even

Above: A raised beach on the island of Arran, Scotland. In areas like this where the sea level has fallen since the end of the last Ice Age, the outline of earlier beaches can clearly be seen, well out of the reach of the highest tides.

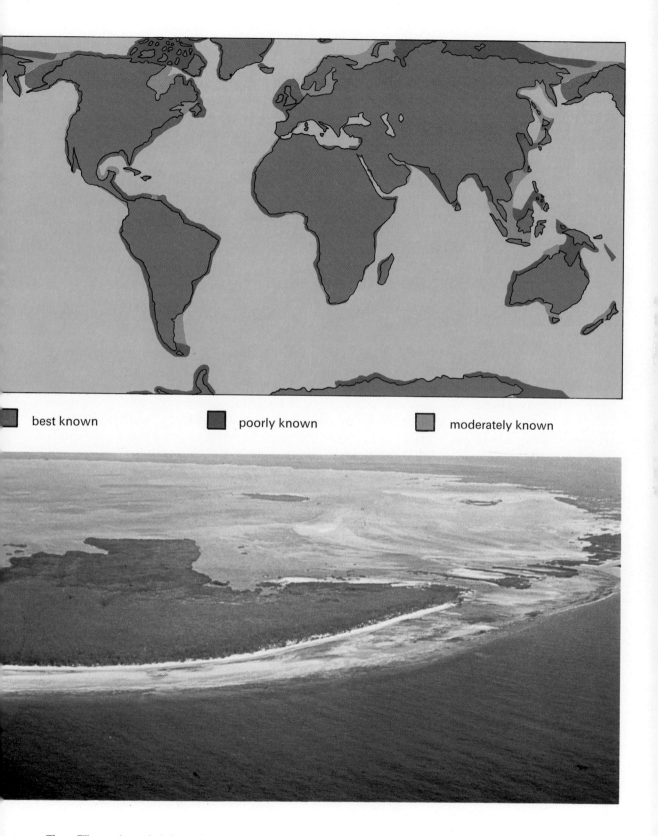

best known

poorly known

moderately known

Top: The continental shelves of the world. Only gradually is a fund of detailed information being accumulated about the shelves, and areas best known are those bordering the countries where scientific knowledge is widely developed and shared.

Above: An aerial view of a coral reef atoll, the island of Aldabra. Offshore the dark blue water shows the sudden steep plunge to the depths of the ocean bottom, unlike the gradual slope of the shelves that surround the world's continents.

boulders, that it can carry along. There is a fairly simple formula that connects speed of flow with *load-potential* (the load a river is capable of carrying). This formula states that the load-potential varies as the sixth power of the speed. In other words, if the speed doubles, the load-potential increases 2^6 times (*i.e.* 64 times), and if the speed trebles, the load-potential increases 3^6 times (*i.e.* 729 times). But as a river nears the sea, its gradient levels out, and it slows down. Then, of course, just the opposite occurs, and the load-potential decreases enormously. The river therefore ceases to roll part of its bed load along. First to be left behind, motionless on the riverbed, are boulders and large stones. Farther along, as the speed slows still more, smaller stones and coarse particles of rock begin to pile up on the riverbed.

Yet even slow-flowing rivers still carry enormous amounts of material, in solution and in suspension, right down to the sea. Each year the world's rivers wash over 8,000 million tons of rock waste into the oceans, some 30 percent of it being in solution. The Mississippi alone removes 500 million tons of solids from the North American land surface annually. And in the region of the Nile Delta, the silt that has been carried down by the river over the years is in places almost 10,000 feet deep. Indeed, in that area the continental shelf is sinking under the great weight, as more and more silt continues to be deposited on it.

When a river reaches the sea, slows down still more, and deposits what is left on its suspended load, many factors combine to prevent that load from simply settling as piles of silt at the river mouth. The sea, with its waves, currents and tides, stirs up the detritus and drags it out onto the continental shelf, to form widely distributed deposits of silt, sand, and mud. These deposits are laid down mostly as shallow-water sediments. Little of the detritus carried down by the rivers reaches the ocean depths.

Much of the dissolved loads that rivers carry to the sea consist of nutrients essential to marine plant life—inorganic nitrates, phosphates, and carbonates that the rains have washed out of the soil and rocks, also organic material from plants and the remains of land animals washed down by the rivers. All these help to replenish the sea's existing store of nutrients.

The most important word in the last sentence is the word "replenish." The sea itself, of course, contains prodigious quantities of organic materials that contain nutrients. The skeletons of innumerable marine animals and plants—such as the coccolithophores, diatoms, and foraminiferans (not to mention those of larger animals)—sink to the seabed and accumulate as oozes. Where they sink to the *deep* ocean bed, however, the nutrients they contain are irretrievably lost. But the general stock of nutrients is topped up to a fairly constant level, largely by the loads that rivers carry down to the continental shelf.

The erosion of land by the rivers, then, enriches the ocean in two ways. It fertilizes the waters of the shelf areas, and at the same time builds up the continental shelves by adding new deposits to them. Moreover, it is not only the bottom that receives a fertilizing dose of debris. Some of the detritus remains in the water, enriching it and creating extremely favorable conditions for bacteria, which are present in great numbers, and for various other kinds of life.

Very occasionally, giant waves called *tsunamis* travel for great distances across the oceans. They are caused by submarine earthquakes or landslides, and are giants among waves. Some tsunamis have caused enormous damage, and on several occasions, great loss of life—some of the most spectacular tsunamis were those caused by the eruptions of the volcano Krakatoa (in the Indian Ocean between Java and Sumatra) on August 26 and 27, 1833. The great waves, at times reaching heights of well over 50 feet, traveled across the Indian Ocean and entered the Atlantic. They were recorded as far north as the English Channel, which they reached only 32$\frac{1}{2}$ hours after the eruption of Krakatoa. In this relatively short space of time, they had traveled 11,040 miles. While at their greatest height, the tsunamis wrought tremendous death and destruction on low-lying coasts and islands of the East Indies; more than 36,000 people were killed.

The normal waves on the surface of the sea are almost entirely due to the wind sweeping over the water. Even when the weather is calm the sea is never quite still. There is almost always a long slow *swell* on the water. Such swells are very long waves generated by storms that may have raged more than 1,000 miles away. They are hardly noticeable in deep water, but in shallow water they turn into visible rollers. The lengths of these swell waves vary greatly, but the average is 500 feet. The effects of waves on the shallow shelf sea is extremely important, because by distributing detritus, waves bring food to the shelf animals. So if we are to understand the conditions of life of shallow water organisms, we need to be clear about the mechanics of wave action.

The wind supplies just about the right amount of energy needed to "move" waves through water. But when we think we see wave crests and troughs "moving," what is actually happening is that the water particles are rotating in very nearly circular orbits, as shown in the diagram on page 118.

The diameter of the orbit is equal to the height of the wave, so that a particle of water is at the top of its orbit at the crest, and at the bottom of its orbit in the trough. It is not only the water at the surface that goes through this circular motion. The water beneath the surface is also orbiting, but the circle it makes decreases rapidly with depth, as shown in the diagram.

The distance from one wave crest to the next is called the *wavelength,* and a particle of water at a depth of half the wavelength follows an orbit whose diameter is only about 1/23rd as great as that of a particle orbiting at the surface. Its speed is also only about 1/23rd as great. Now the average wavelength of swell waves, as we have already seen, is 500 feet. So at a depth of 250 feet, the water in such a swell will be moving 1/23rd as fast as the water at the surface. In practice, this means that halfway down the continental shelf (where the average depth of water is 250 feet) there is still a noticeable

orbital motion from surface waves. And even at the very bottom there is sufficient movement to stir the surface of the sediment. Even more important, this small motion keeps the bottom water in contact with the water higher up, which contains dissolved oxygen. This provides the right conditions for the *aerobic* (oxygen-consuming) bacteria, that live in the bottom sediment and break down detritus into usable nutrients.

However obliquely waves may approach the shore, they usually turn and break more or less parallel to the beach. This is because friction with the shallow bottom slows down their leading edge more than their trailing edge, so that they become straightened out. This is known as *wave refraction*.

Near-shore waves have shorter wavelengths than those farther out to sea. As the waves approach the shore, their crests become closer together. At the same time the wave height increases. As a result, the front of each wave becomes steeper and steeper, until it finally breaks, and the energy it contains is released in the uprush of water on the beach. This is known as the *swash* of the wave. Some of the uprushing water remains on the beach (either trapped in rock pools or filtering through sand and mud) while the remainder, the *backwash*, flows back down the beach into the sea.

On a deeply indented coastline with alternating bays and headlands, the waves advance more rapidly in the deeper water opposite the bays than in the shallower water opposite the headlands. So near the headlands their

Above left: Through shallow water the patterns of deposits of river-borne mud and tide-borne sand show clearly. When a rising tide meets a river, each checks the flow of the other.

Above: Waves of the Atlantic Ocean breaking in even rows on a beach at Angus Point in Ireland. The shallow bottom slows and turns the waves so that they break parallel with the beach itself.

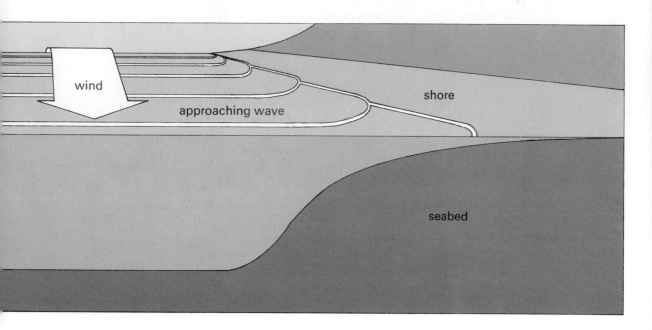

Above: At sea waves are almost entirely created by the wind sweeping over the water. As a wave approaches the shore, the shallow bottom slows down the leading edge more than the trailing edge and the wave tends to turn parallel to the shore.

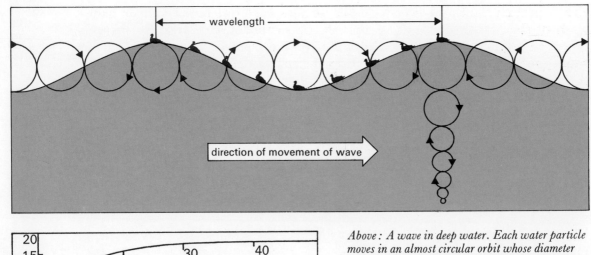

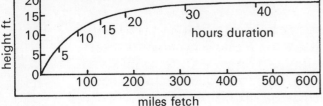

Above: A wave in deep water. Each water particle moves in an almost circular orbit whose diameter at the surface equals the wave height. Below the surface the orbital diameter diminishes sharply. At a depth equal to $^1/_2$ wavelength the diameter is $^1/_{23}$rd that of the diameter at the surface.
Left: The height of a wave in deep water depends on the distance or "fetch" over which a gale has been blowing.

energy is concentrated, and it is there that erosion by wave action and the pounding effect on shore life are greatest.

Not all shores are made of the same material—there are rocky, pebbly, muddy, and sandy shores, with many subtle gradations between all these types. Whether a shore is rocky or made up of pebbles, sand or mud, depends in large part on the action of the waves.

The destruction that breakers wreak on the shores is often underestimated. Almost everywhere and almost every day, they hurl heavy masses of debris-laden water against cliffs, breakwaters, and lighthouses with tremendous force. And where the sea level is rising, waves continually subject more and more of the coastline and drowned river mouths to relentless erosion, on some coasts being aided by rock-boring animals. Wherever there are cliffs, breakers pound at their base, and there the rock is gradually eroded away by two main processes. First, the battering of the waves themselves, and the impact of the stones, sand, and silt that they hurl against the rock, undercut the base of the cliff, until parts of it overhang so much that they collapse. The second process, in which large pieces of even very hard rock may be removed, occurs where a cliff is split along lines of weakness. As the waves pound the cliff face, air pressure builds up in crevices and holes in the rocks. Then, as the waves draw back, the air suddenly expands with an explosive force, driving like a wedge into the rocks.

Erosion is a result of *destructive* wave action. Destructive waves are high, and approach the shore close together, so that each one breaks on the backwash of the one before it. The uprush, or swash, of a destructive wave is therefore reduced by having to travel against the strong backwash. Yet the water particles in the wave are still moving in a nearly circular orbit, and so these waves break with a strong downward force, pounding the shore violently. This loosens material from the shore, which is then carried down to the sea in the strong backwash.

On a shore of solid rock, the first result of long term wave action is to produce a beach consisting mainly of large boulders. These boulders may remain there for a very long time if the rock is hard, or if the coast is but little exposed to strong wave action. However, if the boulders have been broken away from softer rock, or if the wave action is constant and powerful, they will soon be broken up into smaller and smaller pieces. These will then be slowly but steadily ground down against each other to form first rounded pebbles, then sand, and finally silt.

In addition to forming boulder beaches, destructive wave action on a hard rock shore can produce all kinds of dramatic scenic features: wave-cut platforms, caves, and sea-stacks, for example. The undercutting effect of such waves may also carve pedestals or mushroom structures out of the rock.

So far we have considered only the destructive effects of waves. But wave action can also be constructive. This is best seen on *emergent* coastlines— coastlines where the sea level is falling slowly over the years, gradually exposing the gently sloping landward fringe of the continental shelf. The water covering this gentle slope is shallow, and the waves therefore break relatively far out to sea.

Constructive wave action is usually caused by low waves of long wavelength, particularly those ocean-swell waves that have traveled great distances. Because these approach the shore farther apart than destructive waves, the backwash of one wave has time to flow back down the beach before the next wave breaks. The swash, or upwash, of each wave is therefore not held back, and is able to carry stones, silt, and sand up onto the shore. Although the strong swash may carry quite large stones *up* the beach, the rather weak backwash can drag very few back down it. In time this leads to a grading of material. Rocks and large pebbles accumulate at the top of the beach, smaller pebbles lower down, and still smaller ones, together with sand, nearest the water. Where pebbles are piled up steeply at the back of the beach, they tend to roll some way down it under their own weight. Even so, the general gradation pattern still remains.

The materials of beaches are not forever fixed in one geographical spot. They may be carried from one part of the coastline to another by the process called *littoral drift*, which simply means drift along the shoreline. This kind of drifting is controlled mainly by the direction of the prevailing winds, and by tidal streams. Waves driven by strong winds approach the shore at an angle, and therefore carry material obliquely up the beach. As the back-

Top: Waves often cause spectacular erosion along coastlines, as with this tunnel in Tobago (West Indies). Much of this kind of erosion is caused by the abrasive action of stones, sand, and silt that are flung against the cliff by the water.

Above: A cliff that has been undercut by wave action, on the coast of Kenya. If such erosion continues, eventually the cliff edge tumbles down and the waves begin cutting away again. The solitary tree on this beach is a mangrove.

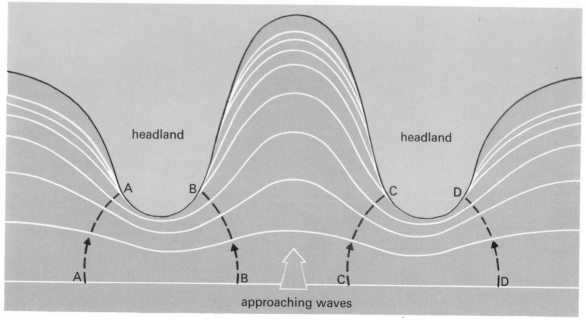

headland

A   B

headland

C   D

A   B   C   D

approaching waves

Top: A wave-cut platform in New Zealand. The grinding action of rocks and sand caught in the waves gradually wears the shore away, producing a platform that has a gentle seaward slope. As the wave erosion continues, the platform broadens.

Above: Wave refraction, which makes waves break nearly parallel with the shore, means that the greatest wave energy is concentrated on the short stretches of shore around a headland (A to B and C to D) and spreads out around a bay (B to C).

wash returns, it describes a very slight curve, but for all practical purposes we can say that it drags material *straight down* the slope of the beach. The net result is that each wave, together with its backwash, carries material *sideways* for a short distance along the beach. Over a prolonged period of time the movement is considerable, and great quantities of small stones, sand, or silt may be transported for long distances.

We have seen how constructive wave action causes a gradation of materials from the back of a beach to the sea line. Littoral drift can cause a similar gradation of size *along* the coastline. As bits of rock are carried farther and farther away from the site where cliff erosion is taking place, they become progressively smaller. There are two reasons for this. First, they rub against each other and become worn down. Second, the smaller and lighter the particles are, the farther they are carried along the coast. So nearest the site of cliff erosion, we find pebble beaches. Next, the beaches are mainly sandy, especially along more sheltered coasts and in bays and inlets. Finally, where there is little wave exposure, the most distant shores are of mud. Along indented coasts subjected to littoral drift, spits and sandbars are built up when some of the coarser sediment being transported is carried straight on and deposited in the deeper water. Thousands of ships have been wrecked on shoals built up in this way.

Along every coast we can see waves. Along most—though not quite all—we are also aware of tides. Tides result from the gravitational pull of the Moon, and to a much smaller extent from that of the far more distant Sun, on the Earth's waters. This pull produces two bulges of water on opposite sides of the Earth. And both these bulges move right around our planet once in about 25 hours—the time the Moon takes to complete one orbit around the Earth.

High and low tides therefore occur twice in every 25 hours—or to be precise, once every 12 hours 25 minutes on average. When high tides are highest and low tides are lowest, they are called *spring tides*. These spring tides occur when the Earth, Moon, and Sun are roughly in a straight line, so that the Sun's gravitational pull is added directly to the Moon's gravitational pull, and this happens at both new and full Moon. (The word "spring" as used here has nothing whatever to do with the season of the year but comes from the Old English word "*springan*," meaning "to rise.") The *neap tides* occur when the pulls of the Sun and Moon on the Earth are at right angles to each other. The neap tides are those with the smallest range—high tides least high, low tides least low. They occur at the first and third quarter of the Moon, when the pull of the Sun is at right angles to the pull of the Moon.

The range of tides also varies, often greatly, from place to place—from a few inches in the Baltic, Mediterranean, and Black Seas to 50 feet in the Bay of Fundy, between Nova Scotia and New Brunswick. Small tidal ranges occur in seas that are almost isolated from oceanic tidal movements. Extremely high ranges occur where oceanic tidal waves run into bottlenecks,

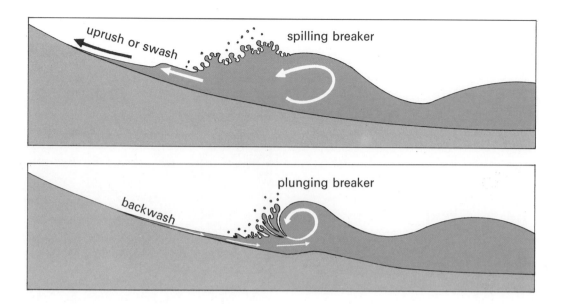

Above: The diagrams show how waves may break on a shore. A spilling breaker, which has a strong forward motion, is actually constructive in its effect on a beach, carrying material up with the swash and leaving most of it behind. A plunging breaker, on the other hand, has a destructive effect, as the strong backwash pulls material from the beach and carries it back into the sea.

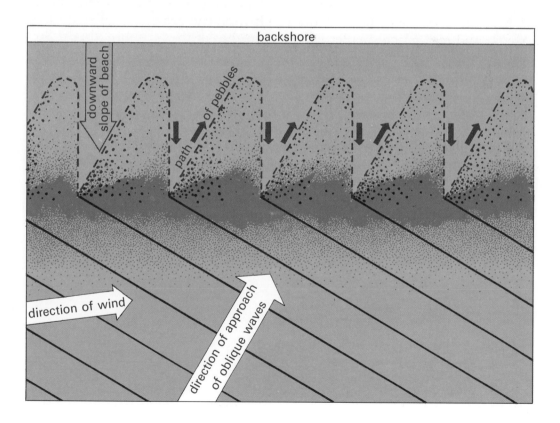

Above: Waves sometimes break at an angle to the shoreline. Then they carry pebbles and sand obliquely up the beach. The backwash, however, carries the particles back down along the slope of the beach. This process is known as littoral drift, and when the direction of the prevailing winds means that most of the time the waves are oblique, huge quantities of materials can be moved for long distances, bit by bit, as each wave carries its load of rocks or sand sideways along the shoreline.

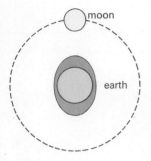

Above: The movement of tides along most coasts is governed by the gravitational pull of the moon and the sun. The sun, although it is vastly larger, is so far away that its pull is much less strong than that of the moon. When the sun and the moon are at right angles to each other, the sun offsets the moon's gravitational attraction, and the range of tides is much smaller. When the sun and the moon are on the same side of the earth, or on opposite sides, their individual effect is strengthened, and the range of tides is greatest.

Right: The Severn Bore in England. A bore is a high tidal wave moving up a river, and the Severn Bore is a particularly spectacular example, with a rush of water as much as four feet high that speeds up the river at up to 12 miles per hour.

so that the water piles up at high tide, and falls a long way below the mean-tide level for the area at low tide. This happens especially in straits and between islands, as well as in narrow sea channels, where the spring tides may have a range of 43 feet. By contrast, the tidal range of mid-ocean is no more than about 16 inches.

Tides and currents are intimately related in narrow seas. Tidal currents may be so powerful that they can dredge shingle and sand along the sea floor, scouring and eroding it in the process. Where rivers reach the sea in estuaries, the shoreward movement generated by the tides is often offset by the more powerful movement of the river out to sea. As currents of highly saline water dip under currents of lighter, less saline water, so in the case of estuaries, the light fresh water of the rivers flows out over the salt water. In this way, the detritus suspended in the river sweeps far out to sea, riding, as it were, on the heavier salt water.

The outer regions of the continental shelf may be somewhat less affected by the discharge of rivers, and by the action of waves and tides, than the near-shore region. But they are also fertile waters, rich in nutrients, because they benefit from the upwelling action of deep currents. As these currents run into the continental slope, they are deflected upward and onto the shelf itself, producing nutrient-rich water. (This can also occur as a result of local wind action, as was noted in considering the Peru and Benguela Currents in Chapter 2.) If we look at a map of western Europe, we can see that from Portugal to the North Cape of Norway and beyond, there is a vast continental shelf standing directly in the way of the North Atlantic Drift section of the Gulf Stream. The whole of this area benefits from the resultant upwelling of nutrients, and much of the history of western Europe is bound up with the bountiful fisheries of this area.

Before discussion of the life of the seashore, and of the whole continental shelf in the next chapter we should note one very important physical condition that affects all life there.

The photic zone of the oceans reaches down to a depth of 650 feet below the surface, which is deeper than the continental shelf. The entire shelf sea lies within the photic zone. More than half of the shelf sea lies within the upper part of the photic zone—the euphotic zone—where all the primary food production by plants takes place. Many of the shelf habitats fall within this zone of primary production, and even the bottom-dwelling animals of the shelf live very close to the productive zone. Life on the shoreward part of the shelf is enriched directly by the nutrients brought down by rivers. And even those creatures farther out are supplied with a thick and steady rain of detritus from above. In the shelf waters there is 30 times as much phytoplankton in proportion to the living space for animals as there is in the waters of the deep oceans. These are the essential reasons why the shelf waters contain such teeming populations of such a wide variety of life.

A small waterfall tumbles down across a beach.
The seashore creatures living near such a source
of fresh water must be adapted to adjust to the
often rapid changes in the levels of salinity,
as the amount of water reaching the beach varies.

7

Life on the Seashore

All along the margins of the sea lies a narrow strip of the earth's surface that belongs to both land and sea. Within this narrow strip are a variety of contours: rugged, rocky headlands, smooth stretches of sandy beach, mangrove swamps, and calm bays. The seashore is subject to the full range of temperatures of day and night, summer and winter; to the full force of breaking waves; to changes in salinity; to the regular ebb and flow of tides; and to drying winds. The animals and plants that live on the seashore must therefore be highly specialized to withstand all these sharp and sudden changes in their environment.

Within a broad habitat, such as the seashore, each species of animal or plant adapted to living under different conditions, often as a result of millions of years of evolution. Biologists call the way of life, or "profession," of the animal or plant its *niche*. In spite of the rigors of life, seashores the world over are among the richest of environments, and they are often densely packed with living things, simply because there are so many different niches to colonize. On many shorelines, every foot of the shore, in every direction, holds something different for the animals and plants that live there.

Yet all over the world more or less the same kinds of animals and plants occupy the same niches. For example, starfish—sometimes even of the same species—are found both in the tropics and in the temperate zone, living in more or less the same way, and on very similar diets. The same goes for crabs, mollusks, seaweeds, and so on.

One reason that we find similar species in such widely separated areas is that the environment in the sea changes little from place to place compared with the environment on the land. Another reason is that many species of marine animals—even those that live attached to the rocks on the seashore—are distributed widely because they have seaborne larvae, which can be carried far and wide by currents and tides.

If you were to walk down the shore toward the edge of the sea, when the tide was out, you would probably notice that the habitats varied progressively down the shore, as did the plants and animals living in them. These

A blenny is eaten by a dahlia anemone, which turns its stomach inside out to digest the fish. In the close-knit community of the shore, each animal species competes ruthlessly with the others for survival. Most are predators, almost all are prey.

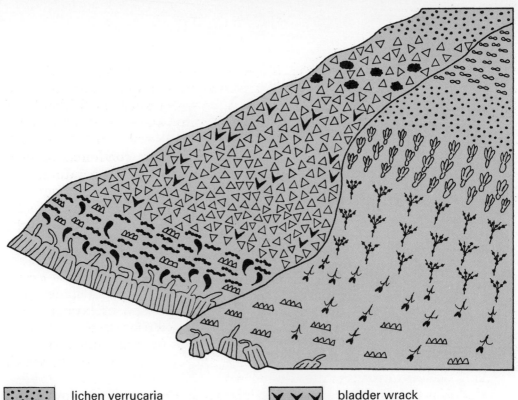

⋮⋮⋮	lichen verrucaria
∞ ∞ ∞	land lichen
⬤⬤ ⬤⬤	lichen lichina
⅄⅄⅄	channeled wrack
Y Y	knotted wrack
❦ ❦	flat wrack

∨ ∨ ∨	bladder wrack
⋀⋀⋀ ⋀⋀⋀	saw wrack
◠	laminarians
◗ ◖	various red seaweeds
～ ～	coralline seaweeds
△ ▽ △	barnacles

In the demanding and constantly changing world of the seashore, each organism has its own niche— the way of life of a plant or animal. In this diagram the arrangement of zones where a specific plant or animal lives is shown for a rocky shore in North Wales. The area to the left is wave-beaten rocks, and the area to the right is very sheltered. The kind of life in each area differs markedly. Although no barnacles are shown on the sheltered area, some species do occur on many such shores.

bands of different living conditions usually run parallel to the coastline, and are called *zones*. If you were on the beach at extreme low tide, you would be able to see the whole of the *littoral zone*, which stretches from the top of the beach down to the water's edge. This region is twice daily submerged and exposed by the tides for variable lengths of time.

Zonation is most easily seen on rocky shores, because the type of ground or *substrate* on or in which the animals live is hard, and few animals can dig into it. On sandy and muddy shores the majority of creatures are hidden, because they bury themselves in the sand or mud. Burrowing animals make up the *infauna* of the seashore. All other animals—those that crawl, glide, or hop on the surface, together with those that remain attached to seaweed, rocks, stones, or other animals—make up the *epifauna*.

At the very top of a rocky shore is a zone that is never covered by the sea, even at high tide, but which is wetted by the spray from powerful breakers. This is the *supralittoral*, or *splash, zone*. On some small islands no part of the land is free from the effects of salt spray, and even on larger land-masses the area influenced by wind-blown spray can stretch quite far inland. Supralittoral zone life must be adapted to withstand conditions of high salinity as well as prolonged drying, rapid changes in temperature, and strong light. But animals living on the surface in the supralittoral zone do not face the problem of being battered by waves. Indeed, they actually benefit from powerful breakers because the farther the spray reaches up the shore, the farther upward they can spread.

Cliff-faces and large boulders at the top of the shore are nearly always encrusted with marine *lichens*, such as the blackish colored *Verrucaria maura*, found on seashores throughout the world. (Lichens are plants made up of single-celled algae and fungi living a closely interdependent existence, the algae producing food by photosynthesis and the fungi drawing up water.) Rocks at the top of the shore are also commonly covered with a slime of simple blue-green algae. These are single-celled plants. (Seaweeds, which can be very large, are all multicelled algae).

Animals of the supralittoral zone include some that have come down from the land toward the sea, and after a long process of evolution of several thousand years, have been able to colonize new niches at the top of the shore. They also include others whose ancestors were once fully marine, but gradually became adapted to life on land. The many immigrants from the land include beetles, flies, and spiders. Other animals typical of the supralittoral zone, on temperate and tropical shores alike, are periwinkles, crabs and sea roaches (known in Britain as sea-slaters). The sea roaches, which are crustaceans, like crabs and shrimps, still have to return to the sea from time to time to wet their gills.

The middle part of the shore is called the *eulittoral zone*. It is usually the widest zone, and every day it is alternately exposed to the air and submerged by the tides. On temperate rocky shores there is often a heavy growth of large seaweeds in the eulittoral zone. These seaweeds show a marked zonation, different species growing at different levels of the shore. In contrast, the seaweeds found on tropical shores tend to be small and inconspicuous, forming a fuzz-like cover on the rocks, and they are not zoned. Even in temperate regions, there may be little growth of seaweeds if the shore is exposed to regular and heavy wave action. On such shores the *sporelings* (the young stages of the seaweeds) do not get a chance to establish themselves on the rocks before they are washed away. But wherever they do have the opportunity of becoming established, many seaweeds are able to withstand considerable battering by the sea. Each plant is anchored firmly to the rock surface by a leathery, multipronged *holdfast* whose tenacious grip is often more than a strong man can break. The stem and fronds are very flexible so that the whole plant can bend with the waves and is not easily torn to pieces. The size to which such a seaweed can grow on the seashore, however, is limited by the strength of its holdfast and its ability to withstand the force of breaking waves.

On the Atlantic coast of Europe, at the top of the eulittoral zone, lives a greenish-yellow colored seaweed known as the channeled wrack. It is exposed to the air for 70–90 per cent of the time, and is well-adapted to withstand drying. Its branched fronds are curled at their edges to form channels—the channels from which it gets its name. When the tide goes out, each channel is left full of water, so that the plant can survive in spite of long exposure to drying winds.

Lower down a typical north temperate shore, if it is sheltered, we may find different kinds of seaweed. These usually occupy quite distinct zones, as shown in the picture below. Some species have bladder-like swellings on their fronds, filled with gas to give them buoyancy. This allows the plants to float with the tide at the surface of the water and trap as much light energy as possible.

The lowest of the littoral zones is called the *sublittoral fringe*. There, life is submerged most of the time, being exposed for only an hour or two on a few days of each month, at spring tides. Here, too, seaweeds dominate. The

Above left: A sea slater rests on a rock at the top of the shore among the seaweed called channeled wrack. The sea slater lives in this splash zone and has evolved over the ages from an ancestor that was a fully marine animal.

Above right: Even within a small area there is a definite zonation of life. Here three brown seaweeds grow closely together. At the top is knotted wrack, to the left at the center is bladder wrack, and saw wrack below.

most common seaweeds of the sublittoral fringe in temperate and cold regions are the large brown seaweeds known as laminarians or kelps. These are giants among the algae, commonly growing to as much as 50 feet long and over 3 feet wide. The laminarian pastures extend beyond the sublittoral fringe into the perpetually submerged *sublittoral zone*.

The seaweeds commonly found along rocky seashores belong to four main groups. These are the blue-green algae, which form a slimy film on the rocks at the top of the shore; the green seaweeds, such as the common sea lettuce being nibbled by the sea hare in the picture on page 135; the brown seaweeds, such as the various species of wracks and the laminarians; and the red seaweeds, such as the delicate, paper-thin dulse, which is gathered for food along the Atlantic coasts of Europe and North America.

As well as the photosynthetic pigment chlorophyll, each of these types of seaweed contains one or more other pigments. It is these additional pigments that give seaweeds of the different groups their characteristic colors. The additional pigments enable the seaweeds to use light of wavelengths that cannot be absorbed by chlorophyll. The red seaweeds, for instance, are often adapted to live beneath the waves in the sublittoral zone, where the light is dim and mainly from the blue and blue-green parts of the spectrum. Their additional pigment, the reddish colored *phycoerythrin*, can absorb energy at

*Above left: Oarweeds (*Laminaria*), usually under water along rocky shores, are exposed at the low water of spring tides. Some* Laminaria *seaweeds, also known as kelps, may grow to a length of 100 to 200 feet, but most are smaller.*

Above right: A tangle of oarweed holdfasts cling to the rock. Laminaria digitata, *shown here, is a common oarweed. A wide variety of small sea animals make their homes under the protection of the broad fronds and holdfast struts.*

these wavelengths and is then able to pass on the energy it has trapped to the part containing chlorophyll, where it is used to make sugars by the process of photosynthesis.

A dense growth of seaweeds can form an extremely effective breakwater. With its many fronds and branches, the surface area of a mass of seaweed is enormous, and creates an effective friction brake on the force of the waves. In fact, a seaweed barrier only a few yards thick can almost eliminate a heavy swell. The huge surface area of a seaweed also provides considerable living space for many animals, some of which bear a surprising resemblance—superficially at least—to plants.

Tiny animals called bryozoans, for instance, are often mistaken for minute seaweeds by the casual observer. Also known as moss-animals, they live in colonies made up of many tiny individuals. The colonies of many species of bryozoan, known as seamats, form encrusting growths on seashore rocks. Each colony is strengthened and protected from wave action by an external skeleton. This is sometimes strengthened with lime. Each individual animal lives within a box-like cavity into which it can withdraw whenever danger threatens. The mouth is surrounded by a ring of tentacles, used for sifting microscopic food particles from the water when the tide is in.

Even plants and animals that do not live on seaweed may benefit from its presence, because it can provide a dim shelter in which organisms adapted to low light intensities can find a home. Furthermore, the seaweed fronds provide ideal places for predators to lurk as well as refuges for their prey.

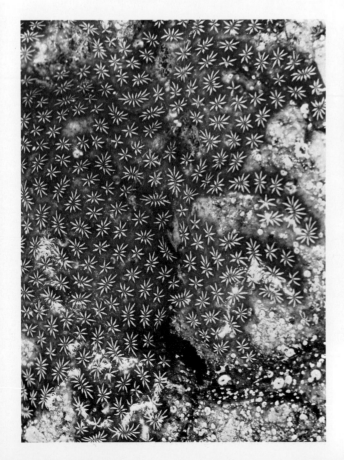

Above: Two acorn barnacles (Balanus crenatus) *feeding, sweeping their fringe-like appendages through the water to trap floating food particles. These barnacles are also in the process of mating.*

Left: The underside of a seashore boulder shows a mass of star-shaped sea-squirt colonies (Botryllus schlosseri). *Each arm of the little star is an individual, with its own mouth, but a second or atrial opening at the base, where they join, is common to the group.*

Above right: Limpets (Patella vulgata) *on a rock at low tide. Although not attached permanently to one place, they can cling to the rock so tightly that the stormiest sea cannot dislodge them.*

Where large seaweeds are unable to gain a foothold, animals must either hide in crevices in or under rocks or in rock pools or live on the exposed rock surface, where they face the dangers of drying out on the upper shore, and damage by waves on the lower shore. Only if they can hide are they able to escape these dangers.

Many of the plant-eating animals of the seashore are grazers. The sea hare, which is a shell-less mollusk and not a relative of the rabbit—eats large seaweeds, and chooses its plant food according to its age. Young sea hares, which are green, feed on the sea lettuce, a bright green seaweed often found in the rock pools of the upper eulittoral zone. And it is not only the fully grown seaweeds that are eaten. Indeed, the microscopic sporelings of new plants are often eaten by sea hares and other grazers as fast as the tiny plants can settle on rocky surfaces.

Throughout the world, the chief grazers on rocky seashores are the limpets. These mollusks—which often look just like animated hats—have settled homes on a particular patch of rock, which they grind down to make an exact fit with their shells. The shell is also ground down in the process, so that the seal is almost perfectly watertight. It is very difficult to dislodge a limpet from its home. Biologists have estimated that to remove a limpet with a base occupying one square inch of the rock surface requires a pull of about 70 pounds! At high water—and sometimes even when completely exposed but in damp cool conditions—the limpet leaves its home and shuffles around, making a clean sweep of all plant growth over an area of up to a square yard.

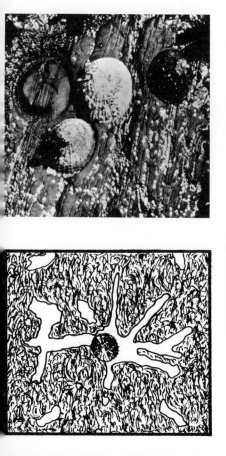

Above: A sea-hare feeding on the seaweed known as sea lettuce. Removed from the water, the sea-hare's soft body collapses into a shapeless mass. Left: A diagram showing the grazing area of a limpet, with the tracks of its feeding sorties.

When the limpet population is reduced for any reason, the seaweeds quickly become luxuriant. Above left: The normal summer appearance of rocks along the Devon coast in southern England. Above right: The same area in July 1967 after detergents were sprayed over the water to remove oil coming from the tanker Torrey Canyon, which had been split open just offshore. Without the pressure of limpets grazing, the seaweeds flourished. A few limpets survived, and the bare patches show their grazing activity. Left: A dog whelk feeding on a group of mussels. Being permanently attached and unable to escape, they are perfect prey for the carnivorous snail.

It scrapes algae off the rocks by using its rasp-like tongue, called the *radula*. It can even deal with large leafy seaweeds. Limpets often eat the plants through at the base, or devour the fronds right through to the midrib. But wherever it roams in its search for food, the limpet always finds its way home before the tide goes out.

Limpets are not the only grazers found on rocky shores. The omnivorous sea urchins, which belong to the large group of spiny-skinned animals called *echinoderms*, are also grazers. And there are several familes of tropical fish that rasp or bite off algae from the rocks or coral reefs.

Many seashore animals are filter-feeders: they filter the water that surrounds them and extract any particles of food it may contain. Animals living in this way can flourish only where certain conditions are met. First there must be a plentiful supply of food particles in the water. Next—because many filter-feeders live fixed in one place—the water must move fast enough to ensure that the food is well distributed. Nowhere are these two conditions better fulfilled than on the seashore. Moreover, the seashore provides a wide range of habitats for animals that live fixed in one place. There they can make their homes on rocks, on the shells of other

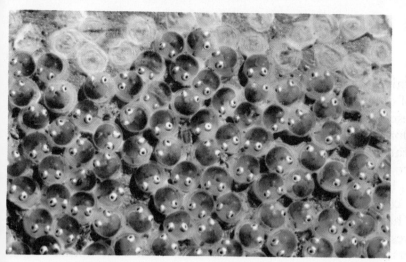

Above: A shanny or blenny, Blennius pholis. *The blenny has no scales, but has a thick, slimy skin. Left: Blenny eggs, adhering to the sides of rock pools in protected sites.*

animals, among the seaweeds, or even—as tube-worms do—in tubes that they themselves construct from sand and their own body-slime. And they can still get enough to eat because the water brings their food to them.

So the littoral zone supports a tremendous variety of filter-feeders—barnacles, sea squirts, sea anemones, many mollusks (such as cockles, razor-shells, and mussels), various species of worms, and a host of others. Some colonies of filter-feeders are among the most densely packed animal communities on our planet.

Often the most conspicuous and numerous animals on the shore are the filter-feeding barnacles, found on rocky coasts throughout the world. The adults look rather like mollusks, but barnacles are really crustaceans—the large group of animals also containing the crabs, shrimps and lobsters. The free-swimming young ones, the larvae, are indeed crab-like, but they undergo a great transformation when they mature and settle down to a fixed life attached to rocks, the bottoms of ships, wooden jetties, or the supports of piers. Their limbs become fringe-like appendages called *cirri*, which they sweep through the water like tiny nets to trap food particles. Any particles larger than about 1/1000 of an inch in diameter are caught among the fine hairs on special large cirri. Then they are removed by smaller cirri that sort out the particles, rejecting those that are inedible and passing the edible pieces to the mouth, where they are swallowed.

Among the many predators of the littoral zone are various snails, bristle-worms, the truly intertidal fish, and shrimps and crabs. Many of these creatures are also scavengers, eating almost anything.

The carnivorous dog-whelk is a snail that is widely distributed on the Atlantic shores of Europe and the northeastern United States. It attacks other mollusks, particularly limpets, top-shells and periwinkles. Its favorite prey, however, are barnacles and mussels, both permanently anchored to their patch of rock and unable to escape. A group of dog-whelks can rapidly transform a barnacle-covered stretch of shore into a mass of empty shells. The feeding whelk extends its *proboscis* (a long structure rather like an elephant's trunk, with the mouth at the end), inserts it between the plates of barnacle shells or the valves of young mussels, and forces them apart. But the tightly shut valves of older mussels and the stout shells of limpets call for sterner measures. In these cases, the dog-whelk simply sits on top of its prey and slowly but surely bores a narrow hole through the victim's shell. This it does with its toothed radula, probably with the help of corrosive secretions from its own body. Once it has managed to worm its proboscis through the hole and into its victim's body, it rasps out the soft flesh with its radula, then passes it back along the conveyer belt of its proboscis into its throat.

Animals such as sea hares, barnacles and dog-whelks are true seashore creatures—permanent residents of the littoral zone. But quite a number of predators and scavengers that feed on rocky shores have their real homes either on land or in the sea. Visitors from the land include several mammals, such as otters and foxes (especially in winter), and man. There are also many land-based birds that take a heavy toll of various shore animals. Among the predators that hunt on the rocky shore are lobsters, starfish, sea urchins, and various fish of the continental shelf waters. All are more or less marine animals and are mostly restricted to the lower levels of the shore. Some species of starfish and sea urchins, however, may be found in tide pools quite far up the shore.

Starfish are fierce predators of bivalve mollusks—those with a shell made up of two *valves* joined by a hinge, like mussels and oysters. The hungry starfish first grips the two valves, holding fast by the suction of the many tube-feet on the underside of each of its five arms. Once in place, the starfish's arms exert a steady pressure, slowly pulling apart the two valves of its helpless victim. Then the starfish turns its own stomach inside out, poking it between the valves into the mussel's body, and digests its meal. When digestion is complete, the starfish retracts its stomach.

There are two other important ways in which many seashore animals obtain their food. These are by living on or within other animals as parasites, and by feeding on detritus. Parasites include leeches, various *nematode worms* (commonly known as threadworms), and certain highly specialized crustaceans. Other small nematodes, and other crustaceans too, feed on the detritus that accumulates in crevices. These parasites and detritus feeders play a very important part in the natural economy of the seashore.

The fish of rocky shores have become adapted to a very different kind of life from that of the fish of the open sea. Many of them live in rock pools, lurking beneath stones or weed. Most species are so well camouflaged that when they are not moving they are almost invisible to predator and prey alike. Some of these fish, such as the gobies, lumpsuckers, "sea snails," and the cling-fish, have the fins on the underside of their bodies modified to form suckers. With these they cling tenaciously onto the rocks and avoid being swept away by the surge of large breakers. Other shore fish, such as the blenny illustrated on the previous page, grip onto rock crevices by their gill-covers, fins, or mouths. Shore fishes either have no scales at all, or minute ones. They are frequently exposed to the full force of waves, and large scales would easily become damaged. In striking contrast to fish of the open ocean, shore fish guard their eggs fiercely. Some species even scrape out special nest-holes in sand or mud, or choose a well-hidden crevice in the rocks to give the eggs added protection, both from the elements and from hungry predators. Shore fish lay very few eggs compared to fish of the open sea, which do not protect their eggs at all. This is because the chances of a shore fish's egg hatching are far greater than those of the entirely unprotected egg of an open-sea fish. This means that the shore fish needs to lay only a few hundred eggs to ensure the survival of the species, whereas open-sea fish must lay thousands or millions of eggs.

Barnacles and limpets are among the epifauna—the surface-dwellers of the seashore. Certain other kinds of rocky shore animals may belong to either the epifauna or to the infauna (the burrowers). These "halfway-house" creatures avoid the risk of drying out by nestling in crevices in the rock, or in spaces between the shells of a bank of mussels or oysters.

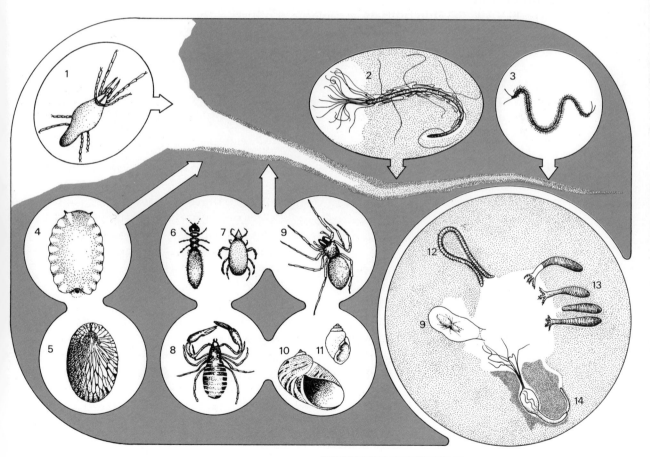

The diagram shows the variety of animals that could be found at various places in a mid-tidal rock crevice. At the narrowest part, only very specialized animals like worms can live, but as the crack gets wider and more exposed, a greater range of small creatures can live in it.

1	red mite	8	pseudoscorpion
2	cirratulid worm	9	marine spider
3	marine centipede	10	trochid snail
4	marine slug	11	colorless snail
5	limpet	12	ragworm
6	springtail	13	peanut worm
7	brown mite	14	terebellid worm

Rocks peppered with the small holes bored by piddocks (Pholas dactylus). Bivalve mollusks, they bore their way into the rock by slowly twisting and rocking against it using the front edge of the shell, which is provided with rows of spines. The opening to a piddock burrow is relatively small, because the animal enlarges the interior of its home as it grows. It has a siphon, which extends to the opening of the burrow. Pholas dactylus has a shell up to six inches long, and is capable of constructing a burrow a foot long. Once it is established, a piddock never moves out of its burrow of its own accord.

Frequently the crevice-dwellers are flattened, and almost all are small, in order to squeeze into narrow spaces.

The true burrowers of rocky shores vary from place to place, according to the hardness of the rock. Soft limestone substrates are inhabited by boring sponges, bristleworms, a variety of bivalve mollusks, and certain sea urchins. Harder rocks have their characteristic infauna too, mainly of bivalve mollusks such as the piddock and rednose. But nowhere on the rocky shore do we find an infauna to compare with that of sandy and muddy shores.

While some rocky shore animals take refuge in crevices or bore into the rock itself, there are many others that live in rock pools. These pools range from large, deep ones to little more than small water-filled crevices in the rock, and they occur from the lowest part of the shore right up to the edge of the splash zone. Rock pools have their supplies of food regularly replenished by the tides, so that the animals living in them do not have to search far for their food. Furthermore, the animals and plants can live in them without facing the danger of drying up—a constant risk to creatures of the exposed shore. But other conditions in a rock pool may fluctuate even more widely and rapidly than they do on the exposed shore, and the animals and plants of the pools must be adapted to cope with these sudden variations. The larger and deeper the pool and the lower its position on the shore, the more its living conditions will resemble those of the offshore waters. Conditions in the pools of the upper shore, however, are less uniform, for these higher pools may be cut off from the sea for days, or even weeks, on end.

The water in the mid-shore pools is replenished twice each 24 hours by the tides. On a hot summer's day the sun quickly warms up the water in these pools, but as soon as the incoming tide floods them with cold water, the temperature drops rapidly. On a cold winter's night the reverse happens. While the tide is out, the water in the pool gets colder than the water in the sea. Then the incoming tide causes a sharp rise in temperature.

Salinity in the pools also changes more drastically than in the rest of the littoral zone. This is because the sun's heat concentrated on a small body of water evaporates it enough to raise the salinity considerably. And the salinity in a well-warmed rock pool will remain high until the next tide, or until there is flooding by rainwater or by inflowing streams. In fact the salinity can change by as much as 30‰ in the space of 24 hours. Rock pool organisms must be able to withstand this rapid variation, and those that can do so are called *euryhaline* species.

There are also variations in the rock pool habitat that are caused by the activities of the plants and animals themselves. During the daytime, while the plants are carrying out photosynthesis, they are using up more carbon dioxide than they give out and giving out more oxygen than they take in. At night, when photosynthesis stops but respiration continues, they are using up oxygen without giving any out, and producing carbon dioxide without taking any in. So each night there is a built-up of carbon dioxide. This dissolves in the water of the rock pool, making it weakly acid. Each day, as the carbon dioxide is used up by the plants, the water becomes more alkaline again. The plants and animals of the rock pools must also be tolerant to these fluctuations.

Many plants and animals have extended their range up the shore by colonizing rock pools, but there are far more dwellers of the upper shore that shun the pools. It cannot be only the rapid fluctuations in physical conditions that keep them out. Animals such as barnacles, and plants such as the channeled wrack, withstand very drastic changes in conditions on the exposed upper shore, yet they do not venture to live in the rock pools. Why this should be so we still do not know.

The teeming life among seaweeds and in crevices or rock pools presents a striking contrast to the apparently lifeless appearance of a sandy or muddy beach. But, as we shall see, this lifelessness is only apparent, for most of the animals of sandy or muddy shores hide themselves even more effectively than do the crevice-dwellers of rocky shores.

There is no sharp distinction between sand and mud, only a difference of particle size (and of organic content, as we shall see later). Particles of coarse sand measure between 1/125 and 1/1250 of an inch in diameter; muddy sand or silt is finer—from 1/1250 to 1/12500 of an inch; and the particles that go to make up mud are even smaller than that. Naturally, not all the particles on a single stretch of beach are of exactly the same diameter, because they are in different states of erosion. It is the *average* particle size that counts.

Early on in the life-history of a sandy beach, constant pounding by breakers reduces all the grains to a fairly uniform size. But once that uniform size is reached, wear and tear ceases, and the sand particles remain the same size almost indefinitely. This is because the smaller a spherical grain becomes, the easier it is for water to form a tenacious film around it. So once the grains are worn down small enough, each one gets a protective water film that separates it from its neighbor and prevents further friction.

The protective film can be destroyed only by drying, and this can happen only when the beach is uncovered at low tide. Even then, evaporation is made extremely difficult by the tenacity of the water film. The result is that—on the lower part of the beach, at least—the sand remains moist almost up to surface level, even at low tide. Anyone who has built sandcastles will know that it is only a matter of digging a few inches below the surface to find sand damp enough to stay put when piled up.

The amount of water held between the particles varies according to their size. On shingle or pebble beaches no water is retained, and these shores are almost lifeless. If the beach is of sand mixed with a good deal of shingle, the water content may be as low as 10 percent of the *wet weight*—the weight of the sand plus the weight of water it holds. On shores made up of finer particles, however, the water makes up about 44 percent of the wet weight, and on some very soft muddy shores, the surface layers may contain over 50 percent water.

Living conditions on soft shores are very different from those on rocky shores. Very few animals or plants are able to live on the surface of the beach because it is constantly churned up by the waves. There is no anchorage or cover save for the occasional stone. A sandy or muddy beach therefore *appears* virtually lifeless at low tide. But such beaches often support huge numbers of animals, nearly all of them living below the surface, either permanently or temporarily, to escape drying when the tide is out. Below the surface it is hardly ever completely dry. Even in the coarse sands on steep beaches, and below the surface of the upper beach (where the water has drained away), the air between the grains is often very moist. In any case, many animals can dig themselves in deeper, until they reach a layer where there is enough moisture for them to survive.

Drying up is not the only hazard from which animals escape when they burrow into sand or mud. At the surface, temperature and salinity fluctuate widely. A few inches below the surface they vary hardly at all. So perhaps it is not surprising that below the surface of sandy beaches—right as far as the back of the beach—we find animals that also make their homes in the sublittoral zone (the zone that is always covered by the shallow sea, and where conditions do not vary greatly). Such animals include many crabs, burrowing sea urchins, and shrimp-like crustaceans.

Sand itself is composed of inorganic matter. But on the beach it is often mixed with varying amounts of organic matter. This comes mostly from decaying seaweed, but also from the feces and dead bodies of animals washed ashore by the tide or brought down by rivers. Such organic matter is extremely important to the life of the seashore, because it provides food for large numbers of small detritus feeders and larger scavengers such as crabs and sea urchins.

Yet too much organic matter on a beach can bring dangers. It can bind the sand or mud grains together, blocking up the spaces between the particles and stopping the circulation of water. If this happens, the deeper layers of sand lose their oxygen, and anaerobic bacteria soon begin to thrive. Most beaches contain a certain amount of iron oxide, and those bacteria change it into iron sulfide, which has a blackish color. A "black layer" is formed that is often foul-smelling, because of the production of hydrogen sulfide gas by the anaerobic bacteria.

On a very sheltered coast with a great deal of organic matter, the black layer may start only a few inches below the surface. But on an exposed wave-lashed coast, where the sand is constantly being churned up, it may be several feet down.

Under these oxygenless conditions, animals living in burrows in the black layer must obtain oxygen for respiration from elsewhere. When there is water overhead they have two main ways of doing so. Some creatures, including certain species of lugworms, draw a current of water through their burrows, and extract oxygen from the water. Others, including masked crabs and cockles, have long respiratory organs that they can poke up through the sand and into the oxygen-rich water. When the surface becomes dry at low tide, animals living in the black layer must either be able to live for a while entirely without oxygen, or else they must be adapted (like some lugworms) to keeping alive on what little oxygen is stored in their blood and tissues.

Few large plants can live on soft beaches because of the lack of firm anchorage. They would simply be washed away by the tides. Where there are some rocks mixed with the sand or mud, we may find a few brown seaweeds near the bottom of the beach and extending out into the shallow sea. On sheltered muddy shores in north temperate regions we may also see various green seaweeds forming loose mats on the surface of the beach. But we will not find anything like the heavy growth of seaweeds that are found where the shoreline is of rock.

Yet in sheltered bays and estuaries in many parts of the world grow a group of flowering plants that have taken to a life in the sea. These sea grasses, as they are commonly called, have true roots that reach deep down into the mud. Although they are known as sea grasses, these plants belong to a completely different group from that containing the grasses of dry land. They are, in fact, related quite closely to the freshwater pondweeds. A good example of the sea grasses is the common eelgrass, which grows to a length of several yards. It is rarely left uncovered by the sea, and at low tide its long, bootlace-like leaves lie waving on the surface of the water.

Until 1930, the eelgrass was very widely distributed in the Northern

Hemisphere, and very plentiful. Danish oceanographers estimate that merely in the Kattegat (the narrow strip of sea between northern Denmark and southern Sweden) no less than 40 million tons of it grew there in 1910. Wherever it grew, its dense waving meadows supported and sheltered a great variety of animals such as small snails, sea hares, cuttlefish and pipefish. And both the eelgrass itself and its inhabitants attracted a variety of seabirds, notably the brent goose, that depended upon it for food.

Today, little remains of the once extensive eelgrass beds of the Northern Hemisphere. The disaster began on the North American Atlantic coast about 1931. Disease struck the plants, disfiguring their leaves with spots and stripes, and causing both leaves and roots to die and rot away. The organism that produced the disease, now believed to have been a marine fungus, was carried by the Gulf Stream to Europe, and there, too, large areas of eelgrass were devastated. A related species, the dwarf eelgrass (which cannot tolerate much immersion, and grows farther up the shore), remained unaffected. Later in the 1930's, the disease spread to Californian coastal waters. There is no possible way in which ocean currents could have carried the infection right around the American landmass, from the Atlantic to the Pacific coast. How, then, did it spread? It is known that at that time young oysters were being transplanted by man from the east coast to the west coast of the United States, and it is probable that they were the carriers of the infection.

Although large plants are rare on soft shores, microscopic plants are usually very abundant. Diatoms and blue-green algae (as well as dinoflagellates and bacteria) are sometimes so numerous that they cover the

Above: Fanworms (Sabella), *spreading their lovely fans, which filter their food out of the water.*
Right: A tubeworm, (Chaetopterus) *that draws water into its tube, filters out food with a mucus film, and stores it in a ball.*

surface of the sand with an oily-looking brown or golden film. And periodically the film may disappear as these minute organisms withdraw into the sand.

Disappearing into the sand or mud is a habit common to many beach creatures, and it often leads people to imagine that a beach is barren when it is really teeming with life. A square foot of sandy beach may look empty enough, but scratch a few inches below the surface and you may find the material for a man-sized meal. For instance, one bivalve mollusk, the European common or edible cockle, often occurs in numbers up to 30 per square foot near the surface. In the Wadden Zee of the Netherlands, its density rises to 185 per square foot. Within the same area, a species of shrimp is known to be taken—by fishes and men between them—at a rate of about 5 per square foot each year, without any apparent decrease in its population.

While plenty of people eat cockles and shrimps, few people would relish a meal of bristleworms. If they did, they could always find a banquet on the beach at La Jolla, California. There, one species of bristleworm has been estimated to have a population density of 3,000 per cubic foot of sand.

Like rocky shores, soft shores harbor a host of filter-feeders—those animals that strain their microscopic diet from the water. Some, such as oysters and mussels, live on the surface of the beach. Others live buried beneath the sand or mud, and must have some means of reaching the water above. When the tide is in, many bivalve mollusks, such as cockles, the razor shells, and Venus shells, live quite near the surface, but at low water they burrow down deeper to keep well-hidden from predators. Their feeding is done when the tide is in—when they can poke their two *siphons* up into

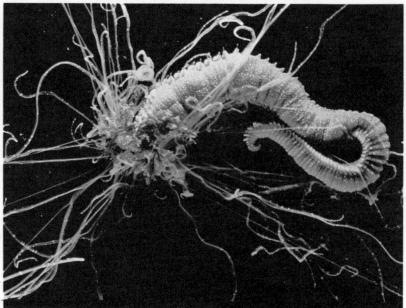

Above: A tubeworm (Amphitrite), *which filters organic matter from sand or mud surfaces. Left: A ragworm* (Nereis), *a predatory, free-living worm that sometimes eats plant and animal debris as well as live prey.*

*Left: A burrowing starfish,
Astopecten irregularis.
Instead of suckers it has pointed
tube feet, adapted for digging.
When placed on sand, the
starfish quickly disappears,
sinking straight down. It feeds
on burrowing mollusks,
swallowing them whole and
disgorging the empty shell or
skeleton later. Below: Methods
of burrowing used by animals
of the soft shore. Soft-bodied
animals, as shown on the left,
force themselves down by using
longitudinal and circular muscles
to elongate and then contract
themselves. Hard-bodied
animals, as on the right, use
specialized appendages to
dig their holes.*

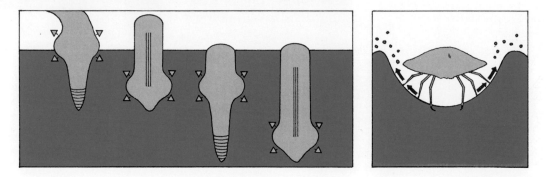

the water. Through the lower *inhalant siphon* water is drawn in and passed through the fine meshwork of the gill chambers, where any food particles are filtered out. The strained-off water, together with waste products and other rejected materials, is passed out through the *exhalant siphon*.

There are other seashore dwellers, called *deposit feeders,* that pick up food particles from the surface of the sand or mud. Some deposit feeders are very choosey about their diet, and are equipped with special lips or tentacles covered with fine hairs called *cilia*, that accept only certain kinds of particles and reject all the rest. Such *selective* deposit feeders include various species of bivalve mollusks, such as telline shells, and bristleworms such as the plumed worm *Amphitrite*. Other deposit feeders are nonselective, living in much the same way as the familiar earthworm does on land. They remain buried and eat the sand or mud wholesale, digesting any organic matter it may contain, and excreting what is useless. An example is the lugworm. When it is dug up (no easy matter, because it can burrow extremely rapidly) it is nearly always found to have its gut full of sand. Nonselective deposit feeders include many worms, shore-living sea cucumbers, and certain mollusks.

Finally there are the predators and scavengers of the sandy or muddy shore. These include ragworms—large active bristleworms with powerful

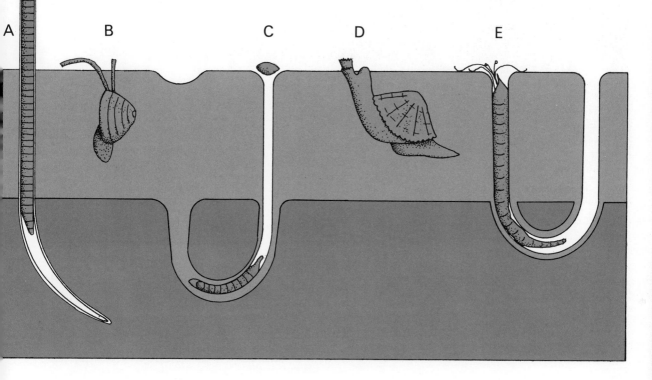

Below: A diagram showing a cross section of some burrows of sand- and mud-living animals. A is a suspension-feeding fanworm Sabella pavonina, *B is the mollusk* Tellina tenuis, *a selective deposit feeder, C is the lugworm* Arenicola marina, *a deposit feeder, D is the cockle* Cardium edule, *a suspension-feeder, and E is the tubeworm* Amphitrite johnstoni, *a selective deposit feeder.*

A   B   C   D   E

jaws. The numerous crabs, too, are either predators or scavengers, and a few carnivorous fish, such as the burrowing sand eels, may also inhabit soft shores. Carnivorous snails, like the dog-whelk of rocky shores, are numerous on soft shores, and are represented by many different species. A common carnivorous snail of sandy Mediterranean shores is *Natica*, which burrows through the sand in search of small bivalve mollusks, feeding on them in the same way as does the dog-whelk.

Many of the burrowing mollusks can make their way very quickly through the sand or mud. Their burrowing tool is a fleshy foot with a pointed tip, which they insert between the sand or mud particles. The end of the foot is then inflated with blood pumped in from the rest of the body. This turns it into a mushroom-shaped bulb, anchoring the animal in the sand or mud. The strong muscles running from the foot to the shell then contract, pulling the body of the mollusk down after it. This process is quickly repeated, and so the animal rapidly buries itself. Some worms, such as the lugworm, burrow by means of a proboscis that can be extended and inflated in much the same way as a mollusk uses its foot. Other worms simply push the sand or mud aside with their pointed heads.

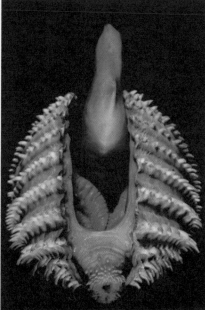

Above left : A male masked crab (Corystes cassivelanus), a burrowing animal found on sandy shores. The crab buries itself in the sand to escape from predators such as the rays, and has a snorkel-like breathing apparatus which it uses when buried. Above right: A cockle (Cardium). It is a filter feeder, with siphons (bottom) used for breathing and feeding. It uses its strong muscular foot (top) for burrowing in soft sand.

Below: Interstitial animals, living between the sand or gravel particles on beaches. On the far left is a female polychaete worm, Sphaerosyllis, with her eggs. Next is the mollusk Microhedyle, then an arthropod, Microcerberus, next the annelid worm Psammodrilus, and at the right a mite. All have become adapted to living in tiny spaces—all are very small, most are elongated and thin, or else flattened vertically or laterally.

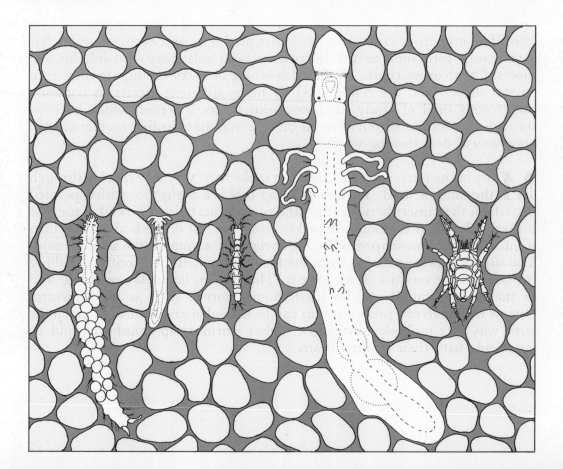

The burrowing habit is not restricted to worms and mollusks. Some of the echinoderms are also adapted for a burrowing life in sand or mud. They include brittle stars, starfish, sea cucumbers, and sea urchins. But all of these are more common in the sublittoral zone than on the seashore.

All the burrowers so far mentioned are relatively large animals. They burrow through the substrate, not as if it consisted of separate particles, but rather as if it were a compact mass. They are therefore found most abundantly in firm beaches of sand or mud, made up of fine particles. But in addition to these larger burrowers, there is a whole world of microscopic animals that live in the spaces *between* the particles. These *interstitial animals,* as they are called by marine biologists, are most abundant on beaches of coarser sand, where there are comparatively large spaces between the grains.

It is only since the 1930's that biologists have begun to realize how common and varied are these tiny creatures. It is now known that they include species from almost every major group of invertebrates, and all of them are well shaped to live in their miniature world. They are mostly very long and very thin, and are thus able to slide easily between the sand grains. In this world, where the space between adjacent grains makes up a home, the single-celled animals called ciliates rank as giants. They can measure as much as 1/25 of an inch in length.

Just like the colossal world of the open oceans, the interstitial world has its own complete food chains. The primary food producers are the diatoms and dinoflagellates which, as we have seen, are often present in such vast numbers as to color the sand. Grazing on these microscopic plants (and on bacteria) are a host of tiny pencil-shaped crustaceans, delicate mollusks, and segmented worms. There are also other kinds of worms, together with slow-moving animals called tardigrades, that pierce the plant walls and suck out the juices from the plant cells. And preying on the grazers—as well as on other minute animals—are various small flatworms and hydroids (animals related to jellyfish and sea anemones).

Mud retains more water between its particles than does sand. This prevents a mud-flat from drying out between tides, even at the surface. What sort of environment does a mud-flat provide? Its softness makes it very easy for burrowers to penetrate, and its permanent moistness ensures that they can avoid drying up without having to dig themselves in deeply. But the slow trickle of water through the mud particles severely limits the entry of oxygen, which therefore becomes scarce at little more than an inch below the surface. And since mud usually contains more organic matter than sand, vast numbers of anaerobic bacteria live in the lower mud layers, and, in the process, produce large quantities of hydrogen sulfide. This accounts for the foul smell of many mud-flats at low tide. An added drawback to life in the mud is that the finely divided particles are liable to choke up the gill systems and filter-feeding apparatus of many animals. On balance, therefore, a mud-flat would not appear to rate highly among marine habitats.

Great numbers of animals do manage to live in mud and muddy sand, nevertheless. There are two ways in which they can do this. One is to stay in the topmost layer only. The other is to live firmly embedded below the oxygenated surface layer, but to take in oxygen—and food—from the surface when the tide is in. Quite a number of mollusks use the second method, employing the "snorkel technique" and using their siphons in much the same way as the filter-feeding cockles of sandy shores. With the inhalant siphon they draw down from above a current of water containing both planktonic food and oxygen. Like the cockles, they are filter-feeders. With the exhalant siphon they discharge water containing waste material. A good example is the clam *Mya arenaria* (popular as a seafood in the United States, where it is known as the soft-shelled clam, and also well known in Europe, where it is known as the gaper). This mollusk digs itself in at a depth of 10 inches. Its shell may be as much as 5 inches long and $2\frac{1}{2}$ inches broad. This is large for a bivalve mollusk, but there are much bigger ones living on the Pacific coast of North America. There the horse clam may reach a length of 8 inches, have siphons over a yard long, and weigh more than 4 pounds. But largest of all such mollusks is the geoduck, which can weigh as much as 11 pounds. It is believed to live for 20 years and can be dug out, with difficulty, from a depth of 6 feet.

Among the animals that use only the top layer of mud-flats are many worms, mollusks, and small crustaceans. They move about on the surface, either browsing on the detritus or swallowing the mud whole and extracting their food from it.

No account of life in the littoral zone would be complete without taking a look at the remarkable animal and plant communities of mangrove swamps, which are such a characteristic feature of many tropical coastlines. There are well-developed mangrove swamps along the shores of Southeast Asia, northern Australia, the Pacific Islands, central America, and Florida. They may develop at the edges of oceanic lagoons, but are more often found in sheltered bays and estuaries, and in them there is a regular zonation of colonizing plants landward, the end result of which may be tropical swamp forest, as in the Florida Everglades.

Mangroves have an extraordinary root system. From the main stem of the tree *prop roots* arch outward and plunge into the mud. From these roots new trees spring up from time to time, so that eventually the whole of a mangrove swamp becomes an impenetrable thicket, except for drainage channels containing tidal water. The mud deposited among the roots is largely anaerobic, but the trees are nevertheless well supplied with oxygen. Growing out vertically from below the mud surface they have slender *pneumatophores*— aerial roots capable of taking in and storing oxygen from the atmosphere.

The high salinity water in which the mangroves live would prevent most plants from drawing up water through their root cells. This is because water tends to pass *out* across the membranes of their root cells from the more dilute cell fluid to the much saltier water outside (by the process of osmosis) and not into the root cells. The cell sap of the mangrove root cells, however, contains a higher concentration of salts than does the water outside and so the plant is able to draw water into them by osmosis.

Beneath the thick canopy of foliage in a mangrove swamp, and among its dense tangle of roots, the air is extremely moist. So animals that normally spend most of their time in water can spend a large part of each day exposed to the air in the mangrove swamps without the risk of drying out. This is particularly important for marine animals, most of which breathe by gills that must remain moist. Mangrove swamps are therefore well suited for marine animals attempting to colonize the land.

Yet despite the wide range of habitats that such swamps offer—leaves, branches, tree trunks, aerial roots, and mud trapped among ground roots—the number of different species living there is not great. This is mainly because the salinity of the swamp changes very considerably with the tides. At high tide, salt water from the open sea covers the lower part of the swamp. Then at low tide, low-salinity water drains down from the land, and tropical rainstorms, too, may flood the area with fresh water. So the animals that live in greatest numbers in mangrove swamps are all *euryhaline*—able to withstand wide variations of salinity. Prominent among them are various species of crabs, mollusks and fish.

Oysters are often found in great numbers, attached to the prop roots of the mangroves by special hook-like projections of their shells. They are eaten both by land animals, such as the racoons of the Florida Everglades, and by sea creatures, such as the crown conch shell.

Fiddler crabs scuttle about on the mud between the mangrove roots at low tide, feeding on plant debris. As soon as the tide comes in, they hurry back to the safety of their burrows, which they dig in the mud. They get their name from the huge claw borne by the male on one of his front limbs—its opposite partner is relatively tiny. He uses the giant claw both for communication and for defense.

Among the strangest creatures of mangrove swamps are the little fish called mudskippers. Various species are found on the coasts of Southeast Asia, India, the Pacific Islands, and West Africa. At low tide they spend most of their time walking over the mud. Some species have suckers on their undersides that help them to climb rocks and mangrove trees. They are on a constant look out for their prey—small crabs, mollusks, worms and insects—with their bulging eyes, borne on swellings on the top of their heads. Each eye can be swiveled independently of the other. They are able to breathe air through modified gills, and can also take in oxygen through their moist skins. Besides walking on their outstretched limb-like front fins, they can also leap like grasshoppers when alarmed or when hunting their prey. This they do by pressing their tails against the mud so that they act like coiled springs when released, sending the little fish flying into the air.

Above: A male fiddler crab, with its immense claw, the "fiddle," which it continually waves around. The fiddler lives among the roots of the mangrove trees in mangrove swamps, taking shelter deep in mud tunnels when the salt tide rises.
Left: A mudskipper, one of the strange small fish found in mangrove swamps, which walks on the land. The pectoral fins are used like hands, strengthened to support the weight of the fish's body out of the water. Spotting its prey with rolling eyes, it feeds on insects, small crabs, and worms.

How do the inhabitants of the seashore react to tidal and seasonal changes? Many animals gear their feeding activities to the regular $12\frac{1}{2}$-hour ebb and flow of the tides. This habit may be inborn, for some of them, including blennies and crabs, have been found to behave in the same way even when they are completely isolated from the shore. Other animals respond to the monthly cycle of spring tides and neap tides, gearing their breeding habits to it. Thus a periwinkle releases planktonic larvae only at spring tides, and huge numbers of a small fish called the Pacific grunion come high up on the sandy shores of California in writhing, slippery masses to lay their eggs. These eggs hatch only at a subsequent spring tide, when the sea once more comes very high up the beach.

In temperate waters, the breeding time is usually related to the seasons of the year, and is aimed to produce young during mild spring or summer weather. Barnacles breed in November, but release their larvae in spring. Certain shore fish, such as blennies, breed in March and their young stay in the rock pools during summer.

Some animals of the seashore do not go through a larval stage. The adults of certain species of periwinkles, blennies, and dog-whelks, for example, lay eggs that hatch directly into creatures like themselves, just as birds do. But a great many animals that will eventually live on the seashore begin life as larvae that are quite unlike the adults, and live for a time in the sea itself. Not only do the larvae that survive find their way to the seashore when they settle down to adult life—they even find their way to suitable spots on the

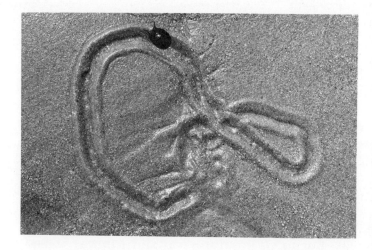

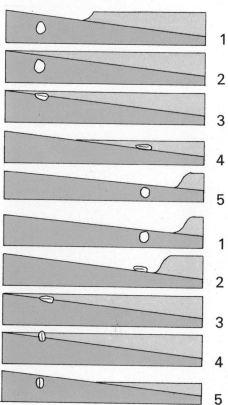

Above: A periwinkle, Littorina littoralis, *when exposed by the ebbing tide keeps its position on the shore by changing the orientation of its movement in relation to the sun. This eventually results in the figure-of-eight path shown here, with the long axis of the eight pointing in the direction of the sun, which it has been following.*

Right: The clam Donax *moves up and down a Japanese beach, migrating with the tide. Triggered by the vibrations of incoming tides, the clam emerges from the sand and is carried farther up the beach. As the wave loses force, it digs in at the new level. Later it emerges to catch the retreating outgoing tide and moves down the beach.*

shore. Experiments have shown that young barnacles and oysters, for example, prefer to settle on areas inhabited by adults of their own species.

Below the level of the lowest spring tides and beyond the influence of vigorous wave action, lies the sublittoral zone. It extends over the whole of the continental shelf, ending where it falls steeply to the abyss.

In the sublittoral zone, which is always covered with water, there is no such zonation as is found on the seashore. Instead, variations in living communities from place to place are mainly of two types, one geographical and the other local. Geographical differences arise from the fact that marine animals, like land animals, are adapted to life at different temperatures. Hence the animals found in the warm inshore waters of the tropics are not the same as those that live in the colder inshore waters of temperate seas. Local variations— especially of the infauna—arise from the fact that the seabed is muddy here, sandy there, and rocky elsewhere. And in different areas of the sublittoral zone, just as on beaches, the kind of burrowing animals present depends on the nature of the substrate.

Before looking at some of the typical animals of the sublittoral zone, it is worthwhile considering its plant life, for it is in this zone that the seaweeds are most impressive. Two factors limit the distribution of seaweeds. One is that they must have rock or boulders to which they can anchor themselves. The other is that they can grow only in the photic zone, where there is enough light for photosynthesis. We have seen that although seaweeds grow in profusion in the rocky littoral zone, the size of a seaweed is limited by the strength of its holdfast, and its resistance to breaking waves. But in the sublittoral zone (where waves do not break) we find that great seaweeds such as *Laminaria* reach their maximum known size—over 300 feet long. Another seaweed, the gigantic *Macrocystis*, also grows both in the littoral and sublittoral zones, but it is most impressive in the latter. It grows as long as 200 feet on the northwest coast of North America, forming dense forests of waving weed, especially in colder waters.

In high latitudes, few seaweeds grow at greater depths than 100 feet. This is probably near the limit for photosynthesis in turbid seas, where the light penetration is not very great. But in clear subtropical and tropical waters, where the sun's rays are more nearly vertical, seaweeds are common at 300 feet, and a few have even been reported from around 600 feet.

The sublittoral zone is rich in animals that live on, or buried in, the sea floor, and each species is adapted to living on or in only one particular kind of substrate. A mud-dweller cannot survive on a rocky bottom, nor a rock-dweller on a mud bottom. Yet a very high proportion of these *benthic*, or bottom-dwelling, animals go through a planktonic or free-swimming larval stage. How do they find their way to the right kind of substrate when the time comes for them to settle down to adult life? During the last few days of their free-swimming life, the larvae test the nature of the substrate, trying to find the kind that suits them. If they cannot find what they need quickly, some of them (including certain species of bristleworms) can delay the changeover to a fixed adult life by as much as a week. The free-swimming larvae may eventually settle on the right kind of substrate if it is already occupied by adult members of their own species. Even so, these larvae have a very high mortality rate. Some fail altogether to find the substrate they need, and die. Countless others are eaten by plankton feeders. It is only because the adults produce so many offspring that each species is able to survive.

The muddy and sandy areas of the sublittoral seabed have animal communities broadly similar to those of corresponding areas of the seashore. Rocky substrates do not usually support a large infauna, but a few bivalve mollusks, such as the date mussel and the giant clam, bore into limestone while the piddock tunnels into a wide range of rocky substrates.

Stony or rocky bottoms have many filter feeders that are either attached to the rock by stalks (like mussels) or cover it like moss (for example, many sponges, sea squirts and the sea mats or bryozoans described earlier). And various carnivores, such as starfish, bristleworms, and certain snails and other mollusks pray on the filter feeders.

The temperate shelf sea is also the habitat of most of the world's edible fish. In general, temperate and cold areas support large populations of comparatively few species, while tropical shelf seas support smaller populations of many more species. And in shelf seas off almost every tropical coast and around tropical islands, rocky substrates bear coral reefs, the most complex and varied animal communities to be found anywhere in the ocean.

There are hundreds of different species of the tiny animals called corals, all belonging to the great group called *coelenterates* (which also includes jellyfish, sea anemones, and hydroids). All the corals have skeletons of calcium carbonate, and many species live together in colonies billions upon billions strong. As they die, their chalky skeletons, huddled closely together, gradually build up great coral reefs from the floors of coastal waters. And these structures are not merely marine graveyards, for on the skeletons of the dead corals, living corals thrive and multiply, adding to the bulk of the reefs.

These reefs are found almost exclusively off shores in and near the tropics—roughly between latitudes 28°N and 28°s—where the inshore waters are almost always warm. Corals do exist outside these warm waters, but it is only there that they form reefs. The right temperature is therefore clearly essential to reef-building. But it is not the only necessary condition. The salinity of the water must also be high enough. Even off warm coasts there are no coral reefs in areas where rivers pour a great deal of fresh water into the sea. Even so, there are areas off the west coasts of Africa and North America where both temperature and salinity seem right for reef-building, but where no reefs exist. This may be because of the upwelling of water in those regions, and possibly also the lack of a suitable substrate. But apart from these somewhat exceptional areas, coral reefs abound in the shallow waters of the tropics, especially at depths of between 100 and 150 feet.

In the Atlantic Ocean there are 35 different species of reef-forming corals. In the Indian and Pacific Oceans the number is no less than 700. Yet the true corals are not the only reef-builders. Two other groups of coelenterates—the hydroids and the *alcyonarians* (soft corals)—also include numerous reef-building species. In addition, there are certain algae that play a part in reef-building, and it is interesting to note that some of the world's earliest-known fossil remains are of such seaweeds.

Many corals live in close interdependence with certain single-celled algae. In the course of photosynthesis, the algae give out oxygen, which the corals need for respiration. In return, the corals give out carbon dioxide, which the algae need for photosynthesis. They also supply the algae with various nutrients essential to healthy growth, such as nitrates and phosphates. However, the algae must have enough light if they are to carry out photosynthesis, and this they can only get in the upper layers of the water. It is this fact that limits the depth at which reefs can flourish.

Like many other marine animals that are fixed to the substrate as adults, corals spend part of their time as free-swimming larvae. Each larva is incubated within the parent coral and, when no bigger than a pin's head, it escapes and swims about by means of tiny hairs or *cilia*. Soon it settles down on a suitable hard surface and gradually changes into the adult form. During the next phase in the coral's life history, it builds up a thick layer of calcium carbonate around itself.

Right: A feather star, Antedon bifida, *a member of the most primitive group of echinoderms. This view is of the underside of the feather star with its pale cirri, used for walking.*
Middle: A brittle star, Ophiocomina nigra, *crawling over a dead razor shell that is encrusted with growths of the small tubeworm* Pomatoceros.
Below left: A scallop, Pecten. *Scallops lie with the shell valves in a horizontal position, rather than vertical as with most bivalves. Each glistening point between the small tentacles is an eye, with lens and retina.*
Below right: The rose "coral" or rose-de-mer Pentapora foliacea, *with small porcelain crabs,* Porcellana longicornis, *sheltering in the crevices.*

Above: A nudibranch, one of the more exotic sublittoral animals. Nudibranches are sea slugs, mollusks that have no shell. They are brightly colored, with plumes that are used for breathing. *Right:* A sea cucumber on the Great Barrier Reef. *Below:* A nudibranch laying eggs on seaweed.

Some corals remain as solitary individuals for the rest of their lives; it is only their non-living skeletons that increase in size. The colonial reef-building corals, however, bud off small, secondary individuals. The form that the coral colony finally takes depends very largely on certain environmental conditions —whether the surrounding water moves rapidly or sluggishly, and the amount of sediments in the water.

Coral reefs and coral lagoons contain the most complex and highly organized living communities in the sea. This is probably due to the unusually high rate of turn-over of nutrients in and around them. It has been calculated that there are more than 3,000 species of animal in Australia's Great Barrier Reef. The coral fish are brightly colored, and some have sharp teeth and elongated snouts so that they can prize out and eat smaller animals lodged in the crevices of the reef. Other reef-dwelling fish (such as parrot fish) browse on the corals themselves, and on algae.

One of the most rapacious predators of the reefs is the crown-of-thorns starfish, which is today causing widespread havoc. A plague of these starfish was recognized in 1959 off northeastern Australia. Since then the species has spread to New Guinea, and to several of the small islands of the Pacific and Indian Oceans. Like all starfish, it feeds very efficiently by pushing its stomach inside out like a glove through the mouth on the underside of its body. And it spreads over the corals, eating them and leaving only the fragile white chalky skeletons behind. It has been calculated that a single crown-of-thorns starfish can destroy 57 square inches of coral surface per day.

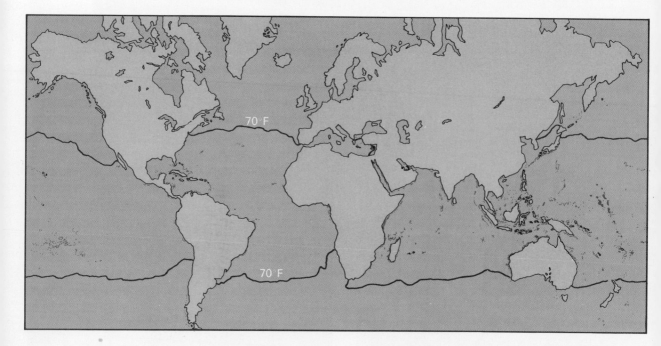

Above: A map of the world showing the distribution of coral reefs. Within the black line, the water temperature is at least 70°F. Corals require a water temperature of over 70°F before they form reefs, although some species can exist in colder water. The reefs are indicated by black dots.

Considered both physically and organically, the reefs are complicated structures, with both the activities of reef-boring animals and the erosion by waves breaking them down as the accumulating calcareous skeletons and the deposition of calcium carbonate from the water build them farther up.

But this starfish has a predator—the giant triton. It is probable that in time the triton will be bred in large numbers and let loose on the reefs in order to control the starfish. Until this program of biological control gets under way, there is a more immediate plan to collect the starfish from the reefs and take it out of harm's way. At the same time the triton is receiving official protection.

Although the recent outbreaks of the crown-of-thorns starfish have done considerable damage to coral reefs in the Pacific and Indian Oceans, they have probably appeared more sensational than they really are because they have been studied in far greater detail than previous outbreaks. Periodic outbreaks of the starfish have occurred in the past, but little was known about the extent of the damage.

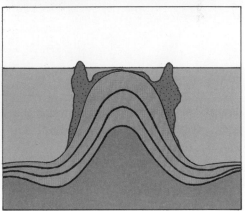

*The stages in the formation of a coral atoll.
Right: An island appears in the ocean, pushed up from the ocean floor by a volcanic eruption.
Middle left: In time, the island sinks slightly, and in the tropics, corals grow in the shallows. As the island sinks farther, the corals deeper than 60 meters die, but new corals form above them.
Middle right: When the island sinks to sea level, a circular atoll is produced, with the corals then growing on the seaward side of the atoll.
Bottom left: After the original island sinks below sea level, the corals begin to form over the top.
Bottom right: Eventually the further subsidence produces a coral island, which is capable of sustaining a variety of plant and animal life.*

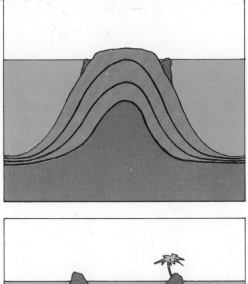

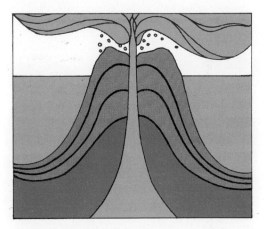

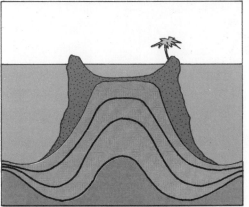

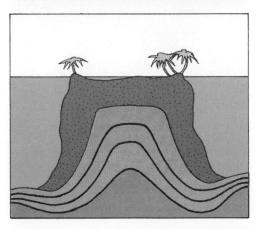

Left: A giant anemone with the fish Amphiprion akallopsis on a coral reef off the coast of Tanzania. Coral reef fish are usually colored vividly, often with fantastic shapes that help them blend into their background. Right: The crown-of-thorns starfish, which is found on the Great Barrier Reef of Australia, and on New Guinea and islands in the Pacific and Indian Oceans. It feeds on coral, destroying over 90% of the corals in some areas of the reef. The most recently described "plague" of starfish, apparently caused by the relaxation of predator pressure, was noticed in 1959. Each starfish is capable of consuming about one square meter of coral per day. Below: Butterfly and damsel fish swim over coral (Pavona). The brilliant colors of the corals are matched by the fish that find refuge in the reef.

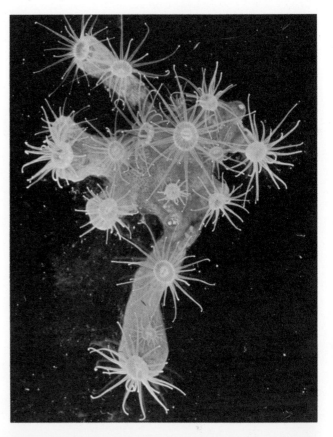

A coral reef is an exotic world, its population varied and colorful, and closely inter-related.
Right: Epizoanthus arenaceus, *a colonial animal resembling an anemone, which habitually grows on other animals, such as sponges and corals.*
Below: A close-up of individual polyps of the red, non-reef forming coral Corallium rubrum.

8

Dominion Over the Sea

A quotation from the Book of Genesis that makes the very first biblical reference to man, goes on at once to couple him with the fish of the sea: "And God said, Let us make man in our image, after our likeness: and let them have dominion over the fish of the sea . . . ." This is not so strange as it may at first seem, for fish have always formed a considerable part of man's diet. Yet, of the 25,000 or so different species of fish those that we eat belong to only a few main groups, and almost all of them live in the waters of the continental shelves.

For centuries the herring and cod families have been the most important fish in human diets. The herring and some of its relatives are among our favorite sea foods, and in 1970 alone the world catch of these fish was almost 9 million tons. During the same period, fish of the cod family provided a world catch of well over 4 million tons.

The cod family includes the cod, the whiting, the haddock and about 15 other species of lesser importance as food. They are all dwellers in the cold and temperate waters of the North Atlantic, where they are found from Cape Hatteras in the south, to Newfoundland, Greenland, Iceland, Norway, and Bear Island (300 miles north of Norway) in the north. They are also plentiful in the North Sea, and in shelf waters as far south as the Bay of Biscay off the coast of Spain.

Study of the cod has shown that at various stages in its life-history it feeds on plankton animals, free-swimming animals, and bottom-living animals. In spring a female cod can lay from 2 to 9 million eggs at one spawning, depending on her size. This huge mass of tiny transparent eggs floats near the surface. The newly hatched young—about $\frac{1}{4}$ inch long— become members of the plankton, feeding on minute copepods. The death toll of the tiny *fry*—as the baby cod are called—is high. Not only are they defenceless against the large predators of the upper waters, but they may also suffer from food shortages.

When the survivors have grown to a little under $\frac{3}{4}$ inch long, they migrate to the seabed. There they feed on larger crustaceans, including small crabs,

The Aswan High Dam, on the upper Nile River near the border of the Sudan. Designed to put an end to the disastrous droughts of Lower Egypt, the dam has also changed the pattern of marine life in both the Red Sea and the Mediterranean waters.

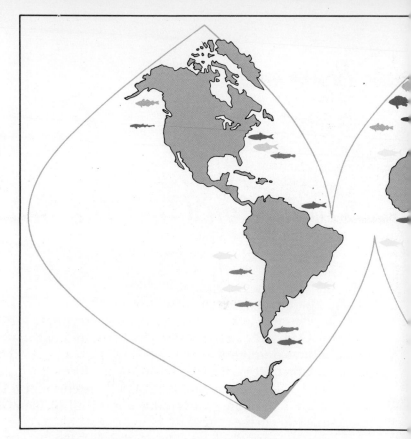

The exploitation of areas in which rich fishing is possible has been greatly expanded in the last 20 years. This map shows the areas that were thought to be underfished in 1949. Today 14 of these areas are probably being fully exploited or in danger of being overfished. The present total ocean harvest is about 55 million metric tons per year. In 1950 it had already increased about 10 times from the harvest a century earlier, in 1850. Even so, it is still true that fish are the only major source of food that is still increasing in global production at a rate that is greater than the rate at which the human population is growing. Therefore the use and intelligent management of ocean resources is becoming a matter of great interest and great concern to the nations of the world, particularly to those that have a tradition of relying heavily on the seas near them both as food and as a source of revenue.

until they reach adulthood at about three years of age. Thereafter they become active predators of other fish, particularly herring, mackerel, and young haddock. But they do not abandon the seabed altogether. Although they are fast and hungry hunters of other fish of the upper waters, cod allow themselves a change of diet from time to time, foraging on the bottom for mollusks, crabs, and bristleworms.

The early history of the haddock is much the same as that of the cod. The young hatch from eggs that look much like cod's eggs and are often found floating with them. And the newly hatched haddock, like the newly hatched cod, join the plankton of the surface waters. However, haddock fry remain there longer than cod fry, and are about 2 inches long before they are ready to journey to the seabed.

The adult haddock develops very different feeding habits from those of the adult cod. It remains on the seabed, browsing for the greater part of the year on worms, crustaceans, mollusks, sea urchins, and brittle stars. During October and November, however, adult haddock feed mainly on the eggs of herrings, which sink to the seabed after being laid.

The haddock's mouth—placed almost on the underside of the head and equipped with fleshy lips and blunt teeth—is well adapted for grubbing about in the seabed. It is in striking contrast to the cod's mouth, which is at the front of the head and armed with strong jaws and sharp teeth that

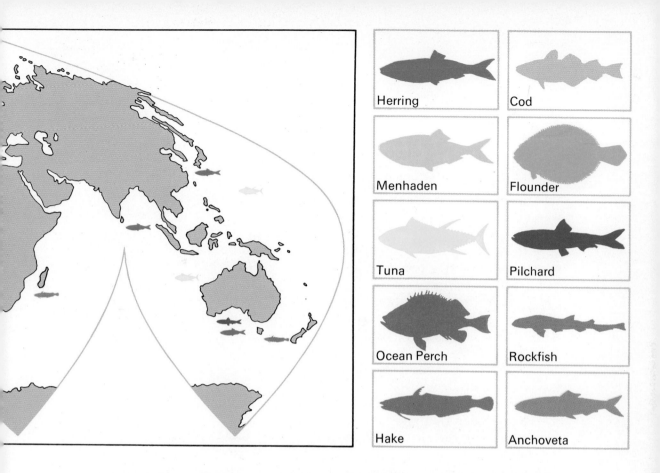

Herring

Cod

Menhaden

Flounder

Tuna

Pilchard

Ocean Perch

Rockfish

Hake

Anchoveta

make it admirably adapted for catching and holding its fast-moving prey.

There is a well-established natural law that if two closely related animals live in the same habitat and occupy the same niche, one of them gradually succeeds at the expense of the other and eventually drives it out. But cod and haddock both thrive in the same habitat. They are able to live together because each makes use of a different niche within that habitat: the cod is the sharp-toothed predator of fast-moving fish, while the haddock with its fleshy mouth roots around in the mud of the seabed for invertebrate food.

While cod and haddock are confined to the North Atlantic, the third important group of food fish, the hake, have a much wider distribution. Various species of hake occur in shelf waters on both sides of the Atlantic, on the Pacific coasts of North and South America, and around New Zealand, and although all these species are closely related, each remains separate from the rest. Cod occasionally travel from east to west across the North Atlantic, but hake species never cross the oceans, nor do they cross the equator. Each species remains on its own continental shelf.

Only in its very early stage does the life history of the hake bear much resemblance to that of the cod or the haddock. The adult female lays about a million eggs at one spawning, and when the young ones hatch out they at first feed on minute copepods. As soon as they are big enough, they graduate to shrimp-like krill until they are two years old. Then they go on to eat other fish, oceanic squids, and even their own young. The stomach contents of adult hake very often consist of 20 percent young hake and 80 percent blue whiting.

Hake have two main habitats. In winter and spring they live beyond the edge of the continental shelf—along the continental slope, to where it reaches a depth of about 2,000 feet. There they feed on whiting and squids. But during the summer and autumn they move shoreward into the shelf waters, and feed near the surface on other fish such as mackerel, horse-mackerel, herring, and garfish. A curious feature of their way of life is that in both their winter and summer habitats they feed only at night, in the surface waters. By day, they rest on the bottom.

All three fishes so far mentioned hatch from floating eggs, yet all of them spend a considerable part of their lives on the seabed. Cod feed there while growing; haddock feed there throughout their adult lives; hake rest there every day, even though they feed near the surface by night.

By contrast, the most important food fish in the Northern Hemisphere, the herring, lays eggs that sink to the bottom. yet the young fish that hatch from these eggs spend most of their lifetime swimming in shoals very near the surface, so near that they are caught mainly in drift nets—nets suspended vertically near the surface of the water from fishing-vessels that drift with the tide. (These are in contrast to the trawlers—fishing boats that move under their own power, dragging their bag-like nets over the seabed to catch fish such as cod, hake, and haddock, which as we saw earlier, are mainly bottom dwellers.)

The North Sea herring lays only about 10,000 eggs at one spawning, as compared with the quarter of a million or more laid by a fully grown haddock, the million or so of a hake, and the 9 million eggs of a large cod. But herring eggs are laid in clumps on pebbly parts of the sea floor and stick to anything solid, even to the backs of crabs. This seems to give them considerable protection. Haddock gorge themselves on herring eggs, and the seabed is teeming with other predators that find both the eggs and young of the herring a tasty meal. Even so, the mortality among young herring must be much lower than that among the fish that lay floating eggs; for although the herring lays relatively few eggs, the sea's herring population is enormous. Herring may spawn during spring, fall, or winter, but whenever they do so, the shoals that congregate at the spawning grounds often contain hundreds of millions of fish.

Soon after they have hatched, the tiny herring larvae swim up toward the surface waters to feed on plant plankton. Later, they switch from this vegetarian diet to become predators of minute crustaceans such as copepods and crab larvae. All this time they are carried along at the mercy of the currents, growing all the while. When they are about $1\frac{3}{4}$ inches long, the young herring mass together in shoals and move from the open sea, where they were hatched, to the coast. There they eat larger planktonic crustaceans, as they do for the rest of their lives. After spending about 6 months in the inshore waters, they return to the open sea. At about $3\frac{1}{2}$ to $4\frac{1}{2}$ years of age, they reach maturity and join up to form huge shoals.

In addition to the edible fish that live in the upper waters, there are some that are specially adapted for life on the seabed. The most important of these is the group known as the flatfish, which include halibut, plaice, flounder, sole, and turbot.

The adult flounder is indeed a flatfish—as flat as if it had been passed through an old-fashioned mangle. Yet the young flounder looks like a "normal" fish. When it is about a month old it begins to undergo an extraordinary change. Gradually the whole of its skull becomes twisted, so that both its eyes come to lie on the animal's right-hand side. So when we see an adult flounder, what appears to be its upper surface is really its flattened

Brittle stars, densely concentrated on the seabed. As bottom-dwelling predators that compete for prey with edible flatfish, such as sole, flounder and plaice, their feeding patterns are important in planning economic harvesting of the oceans.

right side, and what appears to be its lower surface is really its flattened left side. The same is true of an adult halibut, plaice or sole. But with the turbot and a number of other flatfish the left side and not the right faces upward.

Flatfish lay floating eggs—the flounder up to half a million at a time—and these drift in the plankton and develop into tiny larvae that feed on phytoplankton. As they grow, the young fish change to a mixed diet, eating both plant and animal plankton. Then they undergo their remarkable flattening, descend to the bottom, and live on small invertebrates.

Flatfish live in continental shelf waters throughout the world, where the sea floor is in many places crowded with a great variety of other animals, some of which they eat. Since man has an economic interest in flatfish, scientists would obviously like to learn as much as possible about their source of food. But any thoroughgoing study of a population of bottom-living animals presents enormous difficulties, and so far such a detailed study has been made in only a few areas of the world.

For example, in the Kattegat, a partly enclosed branch of the North Sea lying between Denmark and Sweden, Danish marine biologists have been carefully studying the entire seabed for over 40 years. They have found that the total weight of all bottom-dwelling animals in the Kattegat is about 6 million tons. About 5 million tons of these are of no use as food either to the flatfish or to other bottom predators. This leaves approximately 1 million tons of animals (mainly small mollusks, worms, and crustaceans) that can be eaten by all the bottom predators, including the flatfish.

Living on this million-ton food supply are the astonishing figures of 300,000 tons of brittle stars, almost 75,000 tons of other invertebrate predators (crabs, snails, and starfish) and only 5,000 tons of flatfish, most of them plaice. Clearly the brittle stars far outweigh the other predators, and in some areas they appear to cover the seabed to the exclusion of all other living things. But we have to remember that brittle stars are omnivorous creatures, and much of what they eat is mud. Marine biologists generally agree that only about one quarter of their diet consists of animal matter—that is, food for which the other predators are competing. So although there are 300,000 tons of brittle stars in the Kattegat, the weight of them that is competing for the animal food is effectively only a quarter of that value, or 75,000 tons. If we add that weight to the weight of the other invertebrate predators, we can say that there are, in effect, 150,000 tons of invertebrate predators compared with 5,000 tons of flatfish.

The invertebrates and the flatfish feed on the one-million-ton food supply at very different rates. Each invertebrate eats up to 20 percent of its own body weight every day, while a flatfish eats no more than 5 percent of its body weight per day in summer, and very much less in winter. Allowing for all these complications, the flatfish eat no more than about 1 or 2 percent of the total food supply, while the invertebrates eat all the rest.

Similarly, in the Russian White Sea, much of the plankton that would go to feed herring is eaten by jellyfish, which man cannot use as food.

In the Kattegat, as almost everywhere in the sea, there is a delicate balance between all the animals concerned. Sometimes such a balance can be destroyed by a change in conditions. And when this happens it may even affect the course of human history.

Such an event occurred in the Middle Ages. Although it may seem strange that the history of northwest Europe could be altered by the humble herring, it is important to remember that during the Middle Ages, almost all of Europe was Roman Catholic, and the eating of meat was forbidden on Fridays throughout the year as well as the Eve of various important Feast Days, and the 40 days of Lent. The demand for fish was therefore enormous, and it was to a great extent met by the herring fisheries of the Baltic Sea.

The fishing itself was done by Danes, but the trade in salted herrings was completely taken over by the powerful group of north German merchants known as the Hanseatic League. They exported fish to almost every country in Europe, even going to the length of providing their merchant fleets with

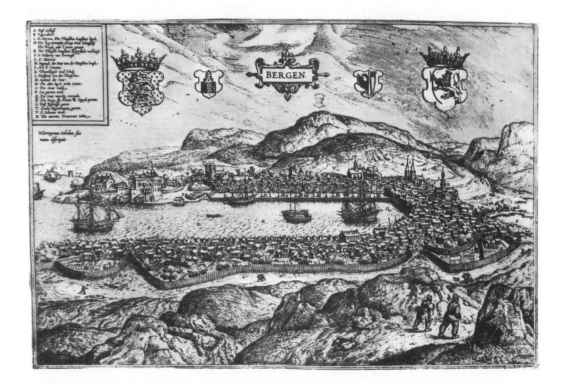

armed convoys to protect them against pirates. As a result, the Hanseatic League became extremely wealthy, and grew to dominate northern Europe from the middle 1200's to somewhat after 1400.

Suddenly, between 1416 and 1425, a major catastrophe hit the Baltic herring. Although the Baltic herring fishery had failed before, and was to fail again in later years, this was the most serious collapse. For some reason, the stocks dwindled to a tiny proportion of the huge shoals of previous years. And when the Baltic fishery collapsed, so did the power of the Hanseatic League.

At this time the Dutch were beginning to fish the herring in the North Sea (which belong to a different race from the Baltic fish), and it was Holland's turn to become a great maritime nation, and to remain so for many years. It has even been said that the foundations of Amsterdam were built on herring-bones. But the best fishing grounds for North Sea herring lay close to the Scottish and English coasts, and this led to disputes, and finally to war, between Holland and Britain. The result was that by 1654 maritime supremacy had passed to Britain.

This historical chain reaction began with the collapse of the Baltic herring fishery. We do not know what happened to the herring, although many reasons have been put forward to explain their disappearance. These include sudden temperature changes and great undersea disturbances that caused a mass migration of the Baltic fish into the North Sea and sealed off their return route.

We do know what is causing the present-day decline in the North Sea herring fishery. Here the reason is all too clear—overfishing by man. By the early years of the present century, the increased efficiency of fishing boats had started to exhaust the stocks of North Sea food fish, particularly the herring. Yet only a few years before, in 1883, the eminent zoologist T. H. Huxley expressed the view that no stocks of commercially fished marine fish would ever be exhausted by fishing.

In recent years, however, fisheries scientists have worked out methods that will ensure a regular supply of fish, after having carefully studied the breeding, growth, and death rates of the fish. These methods involve the taking of only the older fish. The younger, faster-growing individuals are left to replace the depleted stock. These young fish, which were previously caught along with the older ones, now form the main breeding stock. Because they grow faster, the breeding rate increases, more fish are allowed to develop and the old fish that are taken do not seriously deplete the stock.

Although we cannot be certain, it seems very likely that the collapse of the Baltic herring was caused by natural events. The more serious trouble afflicting the Baltic at present, however, is largely the result of man's activities. Today it is not a matter of the decline of a single species, but a major threat to the whole of life in the Baltic.

The flow of fresh water into the Baltic from the many rivers that run into it has always been far greater than the evaporation from its surface. As a result, the heavy high-salinity water tends to stay at the bottom while the light low-salinity water, rich in oxygen, tends to float on top of it. But the Baltic is a very shallow sea, with an average depth of only 180 feet. So until recently the lower layers received enough oxygen to support life at the bottom because of the mixing of the two layers of water during winter storms.

Today, however, the effluent from expanding industries, the sewage of Leningrad, Helsinki, Stockholm and many smaller towns, together with the run-off of agricultural fertilizers from the land, have filled the Baltic with nutrient chemicals. This has led to a great increase in the growth of phytoplankton at the surface. And when all this extra plant life sinks to the bottom there is simply not enough oxygen to enable it to decompose. As a result, the seabed of the Baltic is on its way to becoming lifeless.

Changes in other parts of the world too, are definitely the result of man's activities. In observing such changes, biologists have a rare opportunity to see in a short time something that in nature takes place only over a period of many thousands of years.

Until recent historical times, the Mediterranean and the Red Sea have been wholly separated by land for at least 20 million years. The animal life of the Mediterranean had always had much in common with that of the Atlantic, while marine life in the Red Sea had always been more like that of the Indian Ocean. Just over 100 years ago, in 1869, the Suez Canal was completed, making the first water link between the two seas. One might have thought that it would now be easy for animals to pass from one sea to the other. But for several reasons this did not happen immediately.

The average level of the Red Sea is slightly higher than that of the Mediterranean. So after the canal was opened, there was a slight current from Suez, on the Red Sea coast, to Port Said, on the shores of the Mediterranean. Marine biologists therefore expected that many species of plankton animals and small fish, unable to swim against such a current, would migrate with it from the Red Sea into the Mediterranean. But in 1929, when the canal had been opened for 60 years, only 15 species had done so. None had migrated in the opposite direction.

The reason why so few animals had migrated was that there were two major barriers to hinder them. The first barrier along the length of the Suez Canal was the Bitter Lakes, on the floors of which were enormous accumulations of salt. (The Great Bitter Lake alone contained nearly a billion tons.) So long as the lake was isolated, this salt had remained where it was, more or less capped by a thin layer of very dense, highly saline water. But when the current started to flow through the Suez Canal the salt began to dissolve, and the canal water soon had a salinity of 76–80‰. This proved to be much too high for nearly all Red Sea animals, even though they were accustomed to living in water of relatively high salinity (42‰).

By contrast, the second barrier that migrants had to face was one of *low* salinity. The salinity at Port Said was always lower than that of the rest of

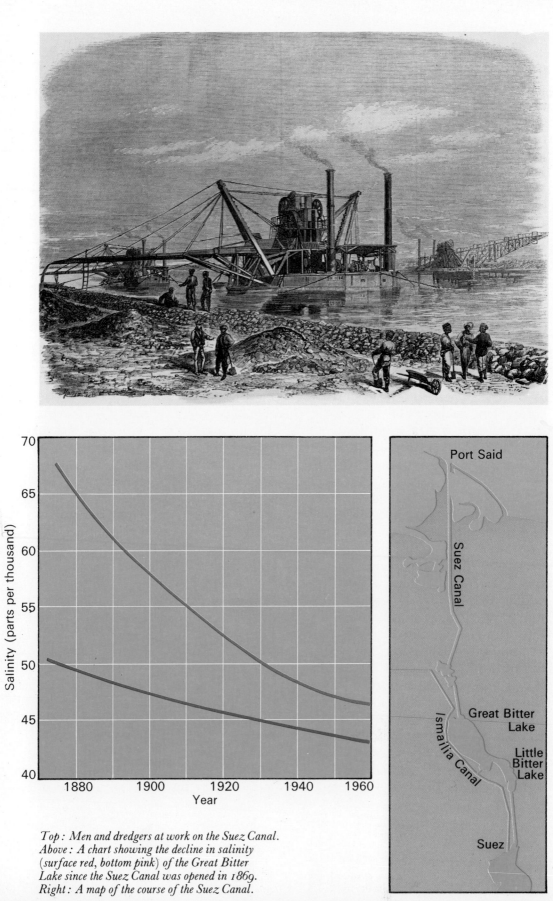

Top: Men and dredgers at work on the Suez Canal.
Above: A chart showing the decline in salinity
(surface red, bottom pink) of the Great Bitter
Lake since the Suez Canal was opened in 1869.
Right: A map of the course of the Suez Canal.

the Mediterranean, because of the immense inflow of fresh water from the Nile River. Indeed, at the height of the Nile flood season the salinity at Port Said fell as low as 26‰. So most of the animals that successfully jumped the salt-water hurdle of the Bitter lakes died at the brackish (low salinity) water hurdle at Port Said.

That was the situation in 1929. Since then the Canal has been deepened and widened several times, and by 1963 the flow had increased by as much as 50 percent. Furthermore, the salt in the Bitter Lakes has now completely dissolved, and much of it has been removed by the canal water, bringing the salinity of the lakes down to 43–46‰—which is close to that of the Red Sea. Thus one of the two barriers—the salt water barrier of the Bitter Lakes—no longer exists. The result has been dramatic. In the last 42 years nine times as many species of animal have migrated along the Suez Canal as did during the first 60 years of its existence. (Nearly all these have been migrants from the Red Sea to the Mediterranean, mainly because of the direction of the current mentioned earlier.)

But the story does not end there. Since the completion of the Aswan High Dam, most of the Nile's flow of fresh water now goes to irrigate Egypt, and no longer reduces the salinity of the sea at Port Said. Thus the second hurdle —the brackish water one—has also been removed. This should result in an even greater migration of animals from the Red Sea to the Mediterranean, just so long as the Canal does not silt up.

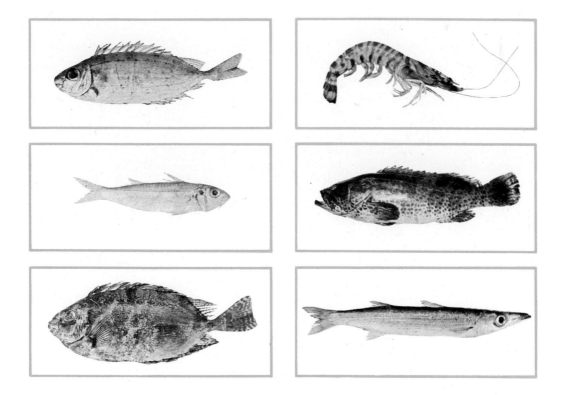

Above: Within a few decades of the opening of the Suez Canal the waters of the Great Bitter Lake had become freshened enough by the Mediterranean to permit migration of these creatures. Top left, rabbit fish, Siganus rivulatus; *top right, prawn,* Penaeus japonicus; *middle left,* Upeneus molloccensis, *related to red mullet; middle right,* Epinephelus tauvina, *grouper; bottom left, rabbit fish,* Siganus luridus; *bottom right, barracuda,* Sphyraena chrysotaenia.

Some of the Red Sea migrants have established themselves so well in the Mediterranean, and are now so plentiful there, that they are fished on a commercial scale. These include a prawn, two species of crabs, and several kinds of fish.

So far the introduction of Red Sea animals into the Mediterranean does not seem to have affected the balance of life there. But sometimes migrants may find themselves in a new environment so much to their advantage that they multiply at a tremendous rate, and upset the delicate balance of the living community into which they have come. It is just this danger that biologists have put forward as a strong objection to building a new, sea-level canal across the Isthmus of Panama. Hardly any marine creatures have been able to migrate across the existing canal, because of its system of locks that lead up on both sides to the fresh water Gatun Lake. But with a new sea-level canal the picture might change very rapidly.

Although the Aswan High Dam has indirectly helped animal migrants from the Red Sea, in another sphere it has brought disaster. For centuries there had been a thriving local sardine fishery centered on the Nile Delta. Sardines require slightly saline water during their larval stages, and they used to be plentiful in the coastal water of the Nile Delta when the great river was in flood and the salinity was low. However, now that the salinity no longer falls, but remains steady at around 39‰—near the normal Mediterranean salinity—the sardines have almost disappeared, and with them the livelihood of the local sardine fishers.

Connected with the Mediterranean by a chain of three comparatively narrow strips of water is a sea of about the same area as the Baltic. This is the Black Sea, with an average depth of 1,300 feet and a maximum depth in the center of over 6,800 feet. Of its total water mass only about one sixth supports animal life; and only about one quarter of the sea floor supports any benthic animals.

The Black Sea, like the Baltic, receives a huge inflow of fresh water from numerous large rivers. Its surface waters therefore have a low salinity (about 18‰) while its deeper waters have a considerably higher salinity. Because the densities are different, there is a little mixing. This leads to a progressive failure in transfer of oxygen from the surface to the deeper water. In the center of the sea the surface waters contain as much oxygen as they can possibly hold in solution, while waters at a depth of 320 feet contain only 5 percent as much as they are capable of holding. Even a little below that depth plankton creatures cannot live; and deeper still the water is completely lifeless (azoic), with an increasing content of dissolved hydrogen sulfide. This *azoic region* forms five sixths of the total water mass. And the entire bottom, at depths greater than about 570 to 650 feet, is also lifeless, except for anaerobic, sulfide-producing bacteria.

Insufficient mixing of surface water with deep water has another serious

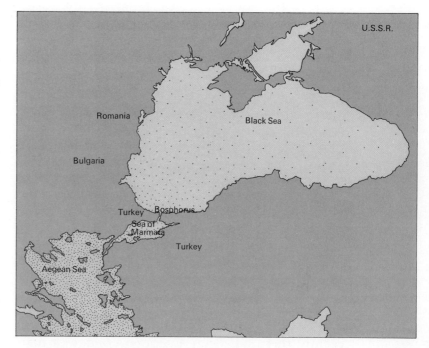

The distribution of species of mollusk in the Balkan seas. Each of the black dots stands for a single species. Moving from the Aegean to the Black Sea through the Sea of Marmara and the Bosphorus Strait the number of species thins, since fewer species can tolerate the lessening salinity of the water and the low winter temperatures.

effect. The Black Sea, like all other seas, receives a steady addition of mineral nutrients, brought down to it by rivers. And, just as in other seas, much of these nutrients sink down to the deep waters, and to the seabed, where there is no plant life to make use of them and bring them back into circulation. But in most sea areas the mixing of waters at different levels brings at least a fair proportion of the nutrients up to the photic zone once more, where growing plants *can* use them. In the Black Sea, however, lack of adequate mixing means that a very high proportion of nutrients simply go on accumulating on the seabed, and are never re-cycled through plants and animals.

The Aegean Sea area of the Mediterranean is connected to the Black Sea by way of the Dardanelles Strait, the Sea of Marmara, and the Bosphorus Strait. The diagram above compares the numbers of species of mollusk in these different areas, and shows how poorly the Black Sea compares with the others in the amount of life it supports.

There are two special conditions that make the Black Sea what it is. The first is the enormous volume of incoming fresh water flowing into it from the Danube, Dnieper, and Don rivers. This fresh water intake is so much greater than the surface evaporation that a strong surface current flows out through the Bosphorus. But the density of the Black Sea as a whole is much less than that of the Aegean, and this leads to the second feature—a saline countercurrent, flowing *into* the Black Sea below the outgoing brackish-water current. It is only this incoming salt water current that prevents the Black Sea from becoming a freshwater lake.

The relationship of salt water to fresh water in the Black Sea has existed

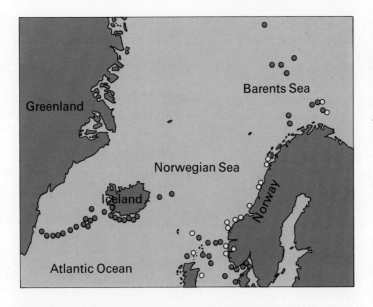

Left: Dumping of industrial and chemical wastes in the oceans can cause havoc, as little is known about the behavior of many chemicals in the sea. This map shows where chlorinated aliphatic hydrocarbons—a by-product of plastics production—were found in the North Atlantic Ocean. The yellow circles mark the samples found in marine life, while the red circles indicate the samples taken from the water.

Right: A western grebe, covered in oil from an oilspill in San Francisco Bay, lies drowned on the bay shore.

ever since the end of the last great Ice Age, about 6,000 years ago. But recently there has been a sudden change of man's making. In an effort to increase the productivity of the land, the Russians are doing an immense amount of dam building and irrigation work on the Dnieper and Don rivers. Similar works are also planned for the Danube. Oceanographers expect that, as a result, the flow of these rivers into the Black Sea will be drastically reduced, and that the salinity of its surface waters will therefore increase. The waters that were formerly of different salinities would then mix, and this mixing would oxygenate the bottom water and make it possible for animal life to exist in the depths. In the long term, the way should then be open to migrants from the Red Sea, which have already reached the Mediterranean, to colonize the Black Sea as well.

While this is the generally accepted view of what will happen, there is a second, much more dangerous, possibility. With surface water and deep water at much the same salinities, there would be far more mixing. Much of the vast accumulation of nutrients now on the seabed would thus rise up into the photic zone, where phytoplankton could make use of them. This could well result in an explosive growth of blue-green algae, as has happened in Lake Erie in recent years; and if that happened there would be a mass slaughter of such life as now exists in the *coastal* waters of the Black Sea region. One can only hope this will not happen. But the mere possibility shows how dangerous it can be to attempt to increase the productivity of the land without full knowledge of what effects such efforts may have on the life of the sea.

"And God said, Let us make man in our image, after our likeness: and let them have dominion over the fish of the sea. . . ." To ensure that his dominion over the fish of the sea benefits him and his posterity to the full, man needs to use every scrap of knowledge and wisdom he can muster.

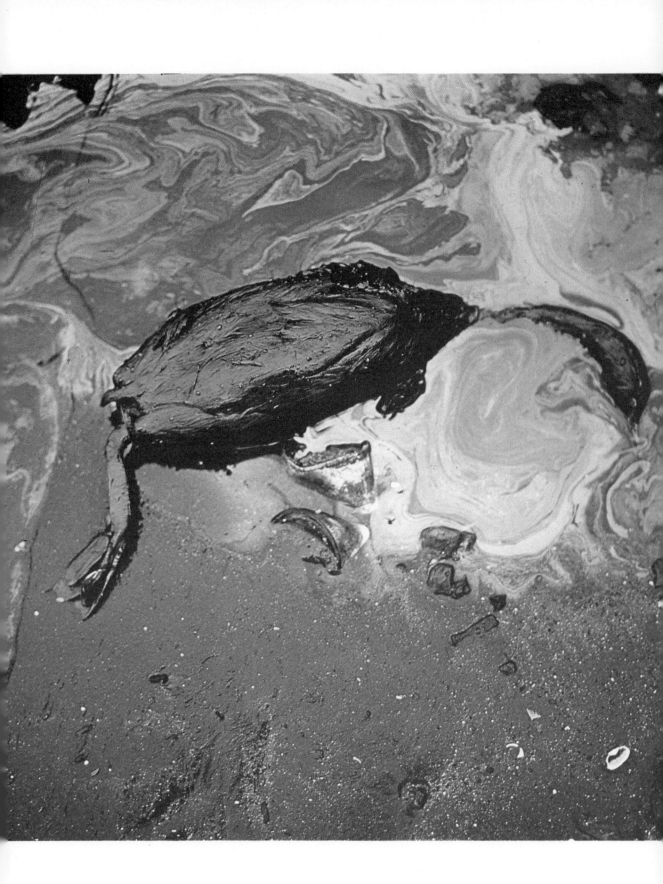

Index

Illustration Credits

80   after F. S. Russell and
C. M. Yonge, *The Seas,*
Frederick Warne & Co.,
Ltd., London

81   Douglas P. Wilson

82   Kelvin Hughes Ltd.

85   British Crown Copyright
Reproduced by
permission of the
Controller of Her
Britannic Majesty's
Stationery Office

87   after Alister Hardy,
Great Waters, Collins,
Publishers, 1967

88   (R) after Alister Hardy in
Discovery Reports, Vol.
XI ,1936, Cambridge
University Press.

89   Peter David

91   after N. A. Mackintosh
in Discovery Reports,
Vol. XV, 1937,
Cambridge University
Press

92   Peter David

96   after N. B. Marshall,
M.A., *Aspects of Deep Sea
Biology,* Hutchinson's
Scientific and Technical
Publications, London,
1954

97   Dr. J. David George,
A.R.P.S.

98   after N. B. Marshall,
M.A., *Aspects of Deep Sea
Biology,* Hutchinson's
Scientific and Technical
Publications, London,
1954

100   (T) Peter J. Green,
Bromley, Kent

102   (T) after photo © Marine
Science Laboratory,
Menai Bridge, Anglesey
(B) after J. R. Norman, *A
History of Fishes,* Ernest
Benn Ltd., London, 1963

104   (T) after N. B. Marshall,
M.A., *Aspects of Deep Sea
Biology,* Hutchinson's
Scientific and Technical

Publications, London,
1954
(B) Popperfoto

105   Peter David

107   after J. R. Norman, *A
History of Fishes,* Ernest
Benn Ltd., London, 1963

108   Popperfoto

109   after N. B. Marshall,
M.A., *Aspects of Deep Sea
Biology,* Hutchinson's
Scientific and Technical
Publications, London,
1964

110   Photo Dennis Brokaw,
San Diego

112   Courtesy J. E. Robinson

113   (B) Dr. J. D. Taylor

116   Photo R. J. Griffith

117   (T) Aerofilms Limited

118   after Morton and Miller,
*The New Zealand Sea
Shore,* Collins,
Publishers, 1968

120   (T) Picturepoint, London
(B) Dr. J. D. Taylor

121   (T) Photo R. J. Griffith
(B) after R. V. Tait,
*Elements of Marine
Ecology,* Butterworth &
Co. (Publishers) Ltd.,
London, 1968

123   (T) after Arthur Holmes,
*Principles of Physical
Geology,* Thomas Nelson
& Sons Limited, 1965
(B) after R. V. Tait,
*Elements of Marine
Ecology,* Butterworth &
Co. (Publishers) Ltd.,
London, 1968

125   Camera Press

127   Keystone

128   Heather Angel, M.Sc.,
F.R.P.S.

130   after A. J. Southward,
D.Sc., *Life on the Sea-
Shore,* Heinemann Edu-
cational Books Ltd.,
London, 1965 (after
Lewis)

132–135 (TL) Heather Angel,

M.Sc., F.R.P.S.

135   (BL) after Hermann
Friedrich, *Marine Biology,*
Sidgwick & Jackson Ltd.,
London, 1969
(R) Heather Angel, M.Sc.,
F.R.P.S.

136   Douglas P. Wilson

137   Marine Science
Laboratory, Menai
Bridge, Anglesey

139   after Morton and Miller,
*The New Zealand Sea
Shore,* Collins, Publishers,
1968

140   Heather Angel, M.Sc.,
F.R.P.S.

144   (L) Dr. J. David George
A.R.P.S.
(R) Heather Angel, M.Sc.,
F.R.P.S.

145   (L) S. G. Giacomelli,
Olmo Arezzo
(R), 146 (T) Heather
Angel, M.Sc., F.R.P.S.

146   (B) after E. R. Trueman,
Nature, 2118, Macmillan
& Co. Ltd., London

147   after A. J. Southward,
D.Sc., *Life on the Sea-
Shore,* Heinemann Edu-
cational Books Ltd.,
London, 1965

148   (TL) Heather Angel,
M.Sc., F.R.P.S.
(TR) Dr. J. D. Taylor
(B) after Morton and
Miller, *The New Zealand
Sea Shore,* Collins,
Publishers, 1968 and
after A. J. Southward,
D.Sc., *Life on the
Sea-Shore,* Heinemann
Educational Books Ltd.,
London, 1965

152   (L) Peter Hill, A.R.P.S.
(R) Jane Burton/Bruce
Coleman Ltd.

153   (L) Heather Angel,
M.Sc., F.R.P.S.
(R) after Joel W.
Hedgpeth, "Sandy

Beaches" in *Treatise on Marine Biology and Paleoecology* Vol. I, ed. Joel W. Hedgpeth, The Geological Society of America, 1957

156 (T) (C) (BR) Heather Angel, M.Sc., F.R.P.S. (BL), 157 Peter Hill, A.R.P.S.

158 after Emery, Tracey and Ladd, in *Treatise on Marine Biology and Paleoecology* Vol. I, ed. Joel W. Hedgpeth, The Geological Society of America

160 Peter Hill, A.R.P.S.
161 (TL) *New Scientist,* London (TR) (B) S. G. Giacomelli, Olmo Arezzo
162 Popperfoto
167 British Crown Copyright Reproduced by permission of Her Britannic Majesty's Stationery Office
169 (T) Staatsbibliothek Berlin (B) The Mansell Collection
172 (T) Mary Evans Picture Library
(BL) after Warren S. Wooster, *The Ocean and Man,* © September, 1969 by Scientific American, Inc. All rights reserved
173 Avinoam Lourie
176 *New Scientist,* London
177 Norman Myers/Bruce Coleman Ltd.

Commissioned Artists:
RUDOLPH BRITTO 28, 61, 80, 88(R), 98, 135(BL), 158
KEVIN CARVER 11, 15, 16-17, 19, 20-21, 24-25, 32-33, 34-35, 37, 54-55, 62(B), 64, 66, 73, 74-75, 78(B), 88(L), 96 102, 117(B), 118, 121(B), 123, 139, 146(B), 148(B), 159, 164-165, 175, 176.
DAVID NASH 47(T), 48, 57, 100 (BL) (BR), 104(T)
DOUGLAS SNEDDON 22, 48-49(B), 87, 91, 94, 107, 109, 113(T), 124(L), 130, 147, 153(R), 172(B)
JOHN TYLER 41(BR), 44, 46, 78(TR), 83, 84